Hypersonics Before the Shuttle

A Concise History of the X-15 Research Airplane

Dennis R. Jenkins

Monographs in Aerospace History
Number 18
June 2000

NASA Publication SP-2000-4518

National Aeronautics and Space Administration
NASA Office of Policy and Plans
NASA History Office
NASA Headquarters
Washington, D.C. 20546

For sale by the U.S. Government Printing Office
Superintendent of Documents, Mail Stop: SSOP, Washington, DC 20402-9328
ISBN 0-16-050363-9

Library of Congress Cataloging-in-Publication Data

Jenkins, Dennis R.
  Hypersonics before the shuttle
  A concise history of the X-15 research airplane / by Dennis R. Jenkins
    p. cm. -- (Monographs in aerospace history ; no. 18) (NASA history series) (NASA
  publication ; SP-2000-4518)
  Includes index.
  1. X-15 (Rocket aircraft)--History. I. Title. II. Series. III. Series: NASA history series
  IV. NASA SP ; 2000-4518.

TL789.8.U6 X553 2000
629.133'38--dc21
                                                          00-038683

Table of Contents

One of the NB-52s flies over the X-15-1 on Edwards Dry Lake in September 1961. (NASA photo EC61-0034)

Preface

Introduction and Author's Comments

It is a beginning. Over forty-five years have elapsed since the X-15 was conceived; 40 since it first flew. And 31 since the program ended. Although it is usually heralded as the most productive flight research program ever undertaken, no serious history has been assembled to capture its design, development, operations, and lessons. This monograph is the first step towards that history.

Not that a great deal has not previously been written about the X-15, because it has. But most of it has been limited to specific aspects of the program; pilot's stories, experiments, lessons-learned, etc. But with the exception of Robert S. Houston's history published by the Wright Air Development Center in 1958, and later included in the Air Force History Office's *Hypersonic Revolution*, no one has attempted to tell the entire story. And the WADC history is taken entirely from the Air Force perspective, with small mention of the other contributors.

In 1954 the X-1 series had just broken Mach 2.5. The aircraft that would become the X-15 was being designed to attain Mach 6, and to fly at the edges of space. It would be accomplished without the use of digital computers, video teleconferencing, the internet, or email. It would, however, come at a terrible financial cost—over 30 times the original estimate.

The X-15 would ultimately exceed all of its original performance goals. Instead of Mach 6 and 250,000 feet, the program would record Mach 6.7 and 354,200 feet. And compared against other research (and even operational) aircraft of the era, the X-15 was remarkably safe. Several pilots would get banged up; Jack McKay seriously so, although he would return from his injuries to fly 22 more X-15 flights. Tragically, Major Michael J. Adams would be killed on Flight 191, the only fatality of the program.

Unfortunately due to the absence of a subsequent hypersonic mission, aeronautical applications of X-15 technology have been few. Given the major advances in materials and computer technology in the 30 years since the end of the flight research program, it is unlikely that many of the actual hardware lessons are still applicable. That being said, the lessons learned from hypersonic modeling, simulation, and the insight gained by being able to evaluate actual X-15 flight research against wind tunnel and predicted results, greatly expanded the confidence of researchers. This allowed the development of Space Shuttle to proceed much smoother than would otherwise have been possible.

In space, however, the X-15 contributed to both Apollo and Space Shuttle. It is interesting to note that when the X-15 was conceived, there were many that believed its space-oriented aspects should be removed from the program since human space travel was postulated to be many decades in the future. Perhaps the major contribution was the final elimination of a spray-on ablator as a possible thermal protection system for Space Shuttle. This would likely have happened in any case as the ceramic tiles and metal shingles were further developed, but the operational problems encountered with the (admittedly brief) experience on X-15A-2 hastened the departure of the ablators.

Many people assisted in the preparation of this monograph. First and foremost are Betty Love, Dill Hunley, and Pete Merlin at the DFRC History Office. Part of this project

Dennis R. Jenkins is an aerospace engineer who spent almost 20 years on the Space Shuttle program for various contractors, and has also spent time on other projects such as the X-33 technology demonstrator.

He is also an author who has written over 20 books on aerospace history.

was assembling a detailed flight log (not part of this monograph), and Betty spent many long hours checking my data and researching to fill holes. I am terribly indebted to her. Correspondence continues with several of the program principals—John V. Becker, Scott Crossfield, Pete Knight, and William Dana. Dr. Roger Launius and Steve Garber at the NASA History Office, and Dr. Richard Hallion, Fred Johnsen, Diana Cornelisse, and Jack Weber all provided excellent support for the project. A. J. Lutz and Ray Wagner at the San Diego Aerospace Museum archives, Tony Landis, Brian Lockett, Jay Miller, and Terry Panopalis also provided tremendous assistance to the project.

Dennis R. Jenkins
Cape Canaveral, Florida
February 2000

With the XLR99 engine lagging behind in its development schedule, the X-15 program decided to press ahead with initial flights using two XLR11 engines—the same basic engine that had powered the Bell X-1 on its first supersonic flight. (San Diego Aerospace Museum Collection)

When the Reaction Motors XLR99 engine finally became available, the X-15 began setting records that would stand until the advent of the Space Shuttle. Unlike the XLR11, which was "throttleable" by igniting different numbers of thrust chambers, the XLR99 was a truly throttleable engine that could tailor its output for each specific mission. (San Diego Aerospace Museum Collection)

Hydraulic lifts were installed in the ramp at the Flight Research Center (now the Dryden Flight Research Center) to lift the X-15 up to the wing pylon on the NB-52 mothership. (Jay Miller Collection)

The early test flights were conducted with a long air data probe protruding from the nose of the X-15. Notice the technician manually retracting the nose landing gear on the X-15, something accomplished after the research airplane was firmly connected to the wing of the NB-52 mothership. (San Diego Aerospace Museum Collection)

Chapter 1

The Genesis of a Research Airplane

It was not until the mid-1940s that it became apparent to aerodynamic researchers in the United States that it might be possible to build a flight vehicle capable of hypersonic speeds. Until that time, propulsion systems capable of generating the thrust required for such vehicles had simply not been considered technically feasible. The large rocket engines that had been developed in Germany during World War II allowed concept studies to be initiated with some hope of success.

Nevertheless, in the immediate post-war period, most researchers believed that hypersonic flight was a domain for unmanned missiles. When an English translation of a technical paper by German scientists Eugen Sänger and Irene Bredt was provided by the U.S. Navy's Bureau of Aeronautics (BuAer) in 1946, this preconception began to change. Expanding upon ideas conceived as early as 1928, Sänger and Bredt had concluded during 1944 that a rocket-powered hypersonic aircraft could be built with only minor advances in technology. The concept of manned aircraft flying at hypersonic speeds was highly stimulating to researchers at the National Advisory Committee for Aeronautics (NACA).[1] But although there were numerous paper studies exploring variations of the Sänger and Bredt proposal in the late 1940s, none bore fruit and no hardware construction was undertaken at that time. It was from this background, however, that the concept for a hypersonic research airplane would emerge.[2]

At the time, there was no established need for a hypersonic aircraft, and it was assumed by many that no operational military[3] or civil requirement for hypersonic vehicles would be forthcoming in the foreseeable future. The need for hypersonic research was not over-whelming, but there was a growing body of opinion that it should be undertaken.

The first substantial official support for hypersonic research came on 24 June 1952 when the NACA Committee on Aerodynamics passed a resolution to "… increase its program dealing with the problems of unmanned and manned flight in the upper stratosphere at altitudes between 12 and 50 miles,[4] and at Mach numbers between 4 and 10." This resolution was ratified by the NACA Executive Committee when it met the following month. A study group consisting of Clinton E. Brown (chairman), William J. O'Sullivan, Jr., and Charles H. Zimmerman was formed on 8 September 1952 at the Langley[5] Aeronautical Laboratory. This group endorsed the feasibility of hypersonic flight and identified structural heating as the single most important technological problem remaining to be solved.

An October 1953 meeting of the Air Force's Scientific Advisory Board (SAB) Aircraft Panel provided additional support for hypersonic research. Chairman Clarke Millikan released a statement declaring that the feasibility of an advanced manned research aircraft "should be looked into." The panel member from Langley, Robert R. Gilruth, played an important role in coordinating a consensus of opinion between the SAB and the NACA.

Contrary to Sänger's conclusions, by 1954 it was generally agreed within the NACA and industry that the potential of hypersonic flight could not be realized without major advances in technology. In particular, the unprecedented problems of aerodynamic heating and high-temperature structures appeared to be so formidable that they were viewed as "barriers" to sustained hypersonic flight.

Fortunately, the successes enjoyed by the second generation X-1s and other high-speed research programs had increased political and philosophical support for a more advanced research aircraft program. The large rocket engines being developed by the long-range missile (ICBM) programs were seen as a way to provide power for a hypersonic research vehicle. It was now agreed that manned hypersonic flight was feasible. Fortunately, at the time there was less emphasis than now on establishing operational requirements prior to conducting basic research, and perhaps even more fortunately, there were no large manned space programs with which to compete for funding. The time was finally right for launching a hypersonic flight research program.[6]

The specific origins of the hypersonic research program occurred during a meeting of the NACA inter-laboratory Research Airplane Panel held in Washington, DC, on 4-5 February 1954. The panel chairman, Hartley A. Soulé, had directed NACA research aircraft activities in the cooperative USAF-NACA program since 1946 and was well versed in the politics and personalities involved. The panel concluded that a wholly new manned research vehicle was needed, and recommended that NACA Headquarters request detailed goals and requirements for such a vehicle from the research laboratories.

In responding to the NACA Headquarters, all of the NACA laboratories set up small *ad hoc* study groups during March 1954. Langley had been an island of hypersonic study since the end of the war and chose to deal with the problem in more depth than the other laboratories. After the new 11-inch hypersonic wind tunnel at Langley became operational in 1947, a research group headed by Charles H. McLellan was formed to conduct limited hypersonic research.[7] This group, which reported to the Chief of the Langley Aero-Physics Division, John V. Becker, provided verification of newly developed hypersonic theories while investigating such important phenomena as hypersonic shock-boundary-layer interaction. The 11-inch tunnel later

served to test preliminary design configurations that led to the final hypersonic aircraft configuration. Langley also organized a parallel exploratory program into materials and structures optimized for hypersonic flight.

Given this, it was not surprising that a team at Langley was largely responsible for defining the early requirements for the new research airplane. The members of the Langley team included Maxim A. Faget in propulsion; Thomas A. Toll in configuration, stability, and control; Norris F. Dow in structures and materials; and James B. Whitten in piloting. All four fell under the direction of Becker. Besides the almost mandatory elements of stability, control, and piloting, a fourth objective was outlined that would come to dominate virtually every other aspect of the aircraft's design— it would be optimized for research into the related fields of high-temperature aerodynamics and high-temperature structures. Thus it would become the first aircraft in which aero-thermo-structural considerations constituted the primary research problem, as well as the primary research objective.

The preliminary specifications for the research aircraft were surprisingly brief: only four pages of requirements, plus six additional pages of supporting data. A new sense of urgency was present: "As the need for the exploratory data is acute because of the rapid advance of the performance of service aircraft, the minimum practical and reliable airplane is required in order that the development and construction time be kept to a minimum."[8] In other versions of the requirements this was made even more specific: "It shall be possible to design and construct the airplane within 3 years."[9] As John Becker subsequently observed, "… it was obviously impossible that the proposed aircraft be in any sense an optimum hypersonic configuration."

In developing the general requirements, the team developed a conceptual research aircraft that served as a model for the eventual X-15. The aircraft they conceived was "… not proposed as a prototype of any of the particular

The first Bell X-2 (46-674) made its initial unpowered glide flight on 5 August 1954. This aircraft made a total of 17 flights before it was lost on 27 September 1956. Its pilot, Air Force Captain Milburn Apt had flown to a record speed 2,094 mph, thereby becoming the first person to exceed Mach 3. (NASA/DFRC)

concepts in vogue in 1954 … [but] rather as a general tool for manned hypersonic flight research, able to penetrate the new regime briefly, safely, and without the burdens, restrictions, and delays imposed by operational requirements other than research." The merits of this approach had been convincingly demonstrated by the successes of the X-1 and other dedicated research aircraft of the late 1940s and early 1950s.[10]

Assuming that the new vehicle would be air launched like the X-1 and X-2, Langley established an aircraft size that could conveniently be carried by a Convair B-36, the largest suitable aircraft available in the inventory. This translated to a gross weight of approximately 30,000 pounds, including 18,000 pounds of fuel and instrumentation.[11] A maximum speed of 4,600 mph and an altitude potential of 400,000 feet were envisioned, with the pilot subjected to approximately 4.5g (an acceleration equal to 4.5 times the force of gravity) at engine burnout.[12]

The proposed maximum speed was more than double that achieved by the X-2, and placed the aircraft in a region where heating was the primary problem associated with structural design, and where very little background information existed. Hypersonic aerodynamics was in its infancy in 1954. The few small hypersonic wind tunnels then in existence had been used almost exclusively for fluid mechanics studies, and they were unable to simulate either the high temperatures or the high Reynolds numbers of actual flight. It was generally believed that these wind tunnels did not produce valid results when applied to a full-scale aircraft. The proposed hypersonic research airplane, it was assumed, would provide a bridge over the huge technological gap that appeared to exist between laboratory experimentation and actual flight.[13]

One aspect of the Langley proposal caused considerable controversy. The Langley team called for two distinct research flight profiles. The first consisted of a variety of constant angle-of-attack, constant altitude, and maneuvering flights to investigate the aerodynamic and thermodynamic characteristics and limitations of then-available technology. These were the essential hypersonic research flights. But the second flight profile was designed to explore some of the problems of manned space flight by making "… long leaps out of the sensible atmosphere." This included investigations into "… high-lift and low-L/D (lift over drag; commonly called a drag coefficient) during the reentry pull-up maneuver" which was recognized as a prime problem for manned space flight from both a heating and piloting perspective.[14]

This brought other concerns: "… As the speed increases, an increasingly large portion of the aircraft's weight is borne by centrifugal force until, at satellite velocity, no aerodynamic lift is needed and the aircraft may be operated completely out of the atmosphere. At these speeds the pilot must be able to function for long periods in a weightless condition, which is of considerable concern from the aeromedical standpoint." By employing a high altitude ballistic trajectory to approximately 250,000 feet, the Langley group expected the pilot would operate in an essentially weightless condition for approximately two minutes. Attitude control was another problem, since traditional aerodynamic control surfaces would be useless at the altitudes proposed for the new aircraft; the dynamic pressure would be less than 1 pound per square foot (psf). The

use of small hydrogen-peroxide thrusters for attitude control was proposed.

While the hypersonic research aspect of the Langley proposal enjoyed virtually unanimous support, it is interesting to note that the space flight aspect was viewed in 1954 with what can best be described as cautious tolerance. There were few who believed that any space flight was imminent, and most believed that manned space flight in particular was many decades in the future, probably not until the 21st century. Several researchers recommended that the space flight research was premature and should be removed from the program. Fortunately, it remained.[15]

Hypersonic stability was the first problem of really major proportion encountered in the study. Serious instability had already been encountered with the X-1 and X-2 at Mach numbers substantially lower than those expected with the proposed hypersonic research aircraft, and it was considered a major challenge to create a solution that would permit stable flight at Mach 7.

Researchers at Langley discovered through

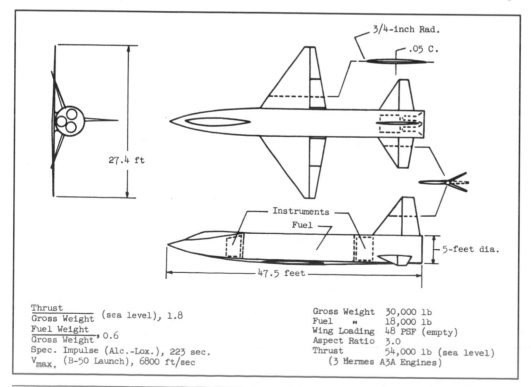

Thrust / Gross Weight (sea level), 1.8
Fuel Weight / Gross Weight, 0.6
Spec. Impulse (Alc.-Lox.), 223 sec.
V_{max}, (B-50 Launch), 6800 ft/sec

Gross Weight   30,000 lb
Fuel   "   18,000 lb
Wing Loading   48 PSF (empty)
Aspect Ratio   3.0
Thrust   54,000 lb (sea level)
    (3 Hermes A3A Engines)

The notional research airplane designed by John V. Becker's group at Langley shows the basis for the eventual X-15. Note the bullet-shaped fuselage (similar to the X-1) and the configuration of the empennage. This was the shape most of the early wind tunnel and analytical studies were performed against. (NASA)

wind tunnel testing and evaluating high speed data from earlier X-planes that an extremely large vertical stabilizer was required if the thin sections then in vogue for supersonic aircraft were used. This was largely because of a rapid loss in the lift-curve slope of thin sections as the Mach number increased. The solution devised by McLellan, based on theoretical considerations of the influence of airfoil shape on normal force characteristics, was to replace the thin supersonic-airfoil section of the vertical stabilizer with a 10 degree wedge shape. Further, a variable-wedge vertical stabilizer was proposed as a means of restoring the lift-curve slope at high speeds, thus permitting much smaller surfaces, which were easier to design structurally and imposed a smaller drag penalty on the airframe. McLellan's calculations indicated that this wedge shape should eliminate the disastrous directional stability decay encountered by the X-1 and X-2.

Becker's group also included speed brakes as part of the vertical stabilizers to reduce the Mach number and heating during reentry. Interestingly, the speed brakes originally proposed by Langley consisted of a split trailing edge, very similar to the one eventually used on the Space Shuttle orbiters. Both the braking effect and the stability derivatives could be varied through wide ranges by variable deflection of the wedge surfaces. The flexibility made possible by variable wedge deflection was thought to be of great value because a primary use of the airplane would be to study stability, control, and handling characteristics through a wide range of speeds and altitudes.[16]

Two basic structural design approaches had been debated since the initiation of the study—first, a conventional low-temperature design of aluminum or stainless steel protected from the high-temperature environment by a layer of assumed insulation; and second, an exposed hot-structure in which no attempt would be made to provide protection, but in which the metal used and the design approach would permit high structural temperatures.[17]

It was found from analysis of the heating projections for various trajectories that the airplane would need to accommodate temperatures of over 2,000 degrees Fahrenheit on the lower surface of the wing. At the time, there

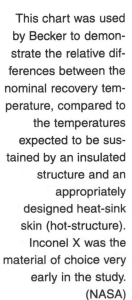

This chart was used by Becker to demonstrate the relative differences between the nominal recovery temperature, compared to the temperatures expected to be sustained by an insulated structure and an appropriately designed heat-sink skin (hot-structure). Inconel X was the material of choice very early in the study. (NASA)

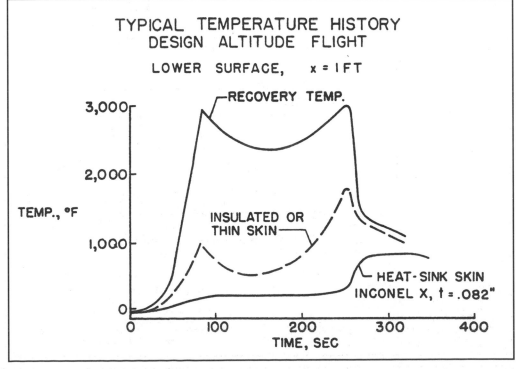

TYPICAL TEMPERATURE HISTORY
DESIGN ALTITUDE FLIGHT
LOWER SURFACE,   x = 1 FT

RECOVERY TEMP.

INSULATED OR THIN SKIN

HEAT-SINK SKIN
INCONEL X, t = .082"

TEMP., °F

TIME, SEC

was no known insulating technique that could meet this requirement. The Bell "double-wall" concept where a non-load-bearing metal sandwich acted as the basic insulator, would later undergo extensive development, but in 1954, it was in an embryonic state and not applicable to the critical nose and leading edge regions. Furthermore, it required a heavy and space-consuming supplemental liquid cooling system. However, the study group felt that the possibility of local failure of any insulation scheme constituted a serious hazard. Finally, the problem of accurately measuring heat-transfer rates—one of the prime objectives of the new research aircraft program—would be substantially more difficult to accomplish with an insulated structure.

At the start of the study it was by no means obvious that the hot-structure approach would prove practical either. The permissible design temperature for the best available material was about 1,200 degrees Fahrenheit, which was far below the estimated equilibrium temperature peak of about 2,000 degrees Fahrenheit. It was clear that some form of heat dissipation would have to be employed—either direct internal cooling or heat absorption into the structure

itself. It was felt that either solution would bring a heavy weight penalty.

The availability of Inconel X[18] and its exceptional strength at extremely high temperatures, made it, almost by default, the structural material preferred by Langley for a hot-structure design. During mid-1954, an analysis of an Inconel X structure was begun by Becker's group; concurrently, a detailed thermal analysis was conducted. A subsequent stress study indicated that the wing skin thickness should range from 0.05 to 0.10 inches—about the same values found necessary for heat absorption in the thermal analysis.

Thus it was possible to solve the structural problem for the transient conditions of a Mach 7 aircraft with no serious weight penalty for heat absorption. This was an unexpected plus for the hot-structure. Together with the fact that none of the perceived difficulties of an insulated-type structure were present, the study group decided in favor of an uninsulated hot-structure design.

Unfortunately, it later proved that the hot-structure had problems of its own, particularly

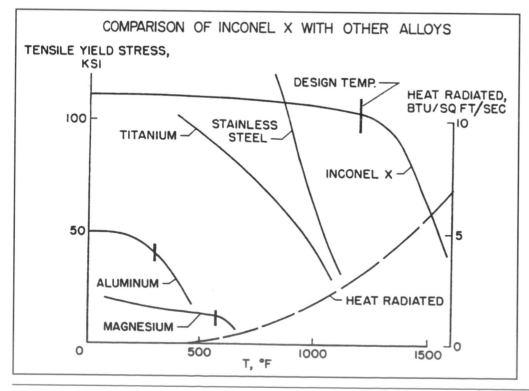

Inconel X was easily the best high-temperature alloy available during the 1950s. It possessed a rare combination of high tensile strength and the ability to withstand high temperatures. Although it proved somewhat difficult to work with, it did not impose some of the problems encountered with titanium on other high-speed aircraft projects. (NASA)

in the area of nonuniform temperature distribution. Detailed thermal analyses revealed that large temperature differences would develop between the upper and lower wing skin during the pull-up portions of certain trajectories. This unequal heating would result in intolerable thermal stresses in a conventional structural design. To solve this new problem, wing shear members were devised which did not offer any resistance to unequal expansion of the wing skins. The wing thus was essentially free to deform both spanwise and chordwise with asymmetrical heating. Although this technique solved the problem of the gross thermal stresses, localized thermal-stress problems still existed in the vicinity of the stringer attachments. The study indicated, however, that proper selection of stringer proportions and spacing would produce an acceptable design free from thermal buckling.

During the Langley studies, it was discovered that differential heating of the wing leading edge produced changes in the natural torsional frequency of the wing unless some sort of flexible expansion joint was incorporated in its design. The hot leading edge expanded faster than the remaining structure, introducing a compression that destabilized the section as a whole and reduced its torsional stiffness. To negate this phenomenon, the leading edge was segmented and flexibly mounted in an attempt to reduce thermally induced buckling and bending.

With its research objectives and structure now essentially determined, the Langley team turned its attention to the questions of propulsion by examining various existing rocket propulsion systems. The most promising configuration was found to be a grouping of four General Electric A1 or A3 Hermes rocket engines, due primarily to the "thrust stepping" (a crude method of modulating, or throttling, the thrust output) option this configuration provided.

The studies prompted the NACA to adopt the official policy that the construction of a manned hypersonic research airplane was fea-

sible. In June 1954, Dr. Hugh L. Dryden sent a letter to Lieutenant General Donald Putt at Air Force Headquarters stating that the NACA was interested in the creation of a new manned research aircraft program that would explore hypersonic speeds and altitudes well in excess of those presently being achieved. The letter also recommended that a meeting between the NACA, Air Force Headquarters, and the Air Force SAB be arranged to discuss the project. Putt responded favorably, and also recommended that the Navy be invited to participate.

NACA representatives met with members of the Air Force and Navy research and development groups on 9 July 1954 to present the proposal for a hypersonic research aircraft as an extension of the existing cooperative research airplane program. It was soon discovered that the Air Force SAB had been making similar proposals to Air Force Headquarters, and that the Office of Naval Research had already contracted with the Douglas Aircraft Company to determine the feasibility of constructing a manned aircraft capable of achieving 1,000,000 feet altitude. Douglas had concluded that 700,000 foot altitudes would be possible from the reentry deceleration standpoint, but that the thermostructural problem had not been thoroughly analyzed. It was agreed that a cooperative program would be more cost effective and likely lead to better research data at an earlier time.[19]

The Navy and Air Force representatives viewed the NACA proposal with favor, although each had some reservations. At the close of the meeting, however, there was agreement that both services would further study the details of the NACA proposal, and that the NACA would take the initiative to secure project approval from the Department of Defense.[20]

Less than a month later, the Air Force identified the principal shortcoming of the original Langley proposal—the apparent lack of a suitable rocket engine. In early August the Power Plant Laboratory at the Wright Air Development Center (WADC) pointed out that "no current rocket engines" entirely satisfied the NACA requirements, and emphasized that the Hermes engine was not designed to be operated in close proximity to humans—that it usually was fired only when shielded by concrete walls. Other major objections to the Hermes engine centered around its relatively early state of development, its limited design life (intended for missile use, it was not required to operate successfully more than once), and the apparent difficulty of incorporating the ability to throttle it during flight.[21] WADC technical personnel who visited Langley on 9 August drew a firm distinction between engines intended for piloted aircraft and those designed for missiles; the NACA immediately recognized the problem, but concluded that although program costs would increase, the initial feasibility estimates would not be affected.[22]

WADC's official reaction to the NACA proposal was submitted to the Air Research and Development Command (ARDC) on 13 August.[23] Colonel V. R. Haugen reported "unanimous" agreement among WADC participants that the proposal was technically feasible; excepting the engine situation, there was no occasion for adverse comment. The evaluation forwarded by Haugen also contained a cost estimate of $12,200,000 "distributed over three to four fiscal years" for two research aircraft and necessary government-furnished equipment. Estimated costs included: $1,500,000 for design work; $9,500,000 for construction and development, including flight test demonstration; $650,000 for government furnished equipment, including engines, $300,000 for design studies and specifications; and $250,000 for modification of a carrier aircraft.[24] Somewhat prophetically, one WADC official commented informally: "Remember the X-3, the X-5, [and] the X-2 overran 200 percent. This project won't get started for $12,000,000."[25]

On 13 September, the ARDC issued an endorsement of the NACA proposal, and recommended that the Air Force "... initiate a project to design, construct, and operate a new

research aircraft similar to that suggested by NACA without delay." The aircraft, emphasized ARDC, should be considered a pure research vehicle and should not be programmed as a weapon system prototype. On 4 October 1954, Brigadier General Benjamin S. Kelsey, Deputy Director of Research and Development at Air Force Headquarters, stated that the project would be a joint Navy-NACA-USAF effort managed by the Air Force and guided by a joint steering committee. Air Force Headquarters further pointed out the necessity for funding a special flight test range as part of the project.[26]

The NACA Committee on Aeronautics met on 5 October 1954 to consider the hypersonic research aircraft. During the meeting, historic and technical data were reviewed by various committee members including Walter C. Williams, De E. Beeler, and research pilot A. Scott Crossfield from the High-Speed Flight Station (HSFS). Williams' support was crucial. Crossfield would later describe Williams as "... the man of the 20th Century who made more U.S. advanced aeronautical and space programs succeed than all the others together. ... He had no peer. None. He was a very strong influence in getting the X-15 program launched in the right direction."[27]

Although one Committee member expressed opposition to the proposed hypersonic research aircraft as an extension to the ongoing test programs, the rest of the Committee supported the project. The Committee formally adopted a resolution to build a Mach 7 research airplane (attached as an appendix to this monograph).[28]

Because the anticipated cost of the project would require support from Department of Defense contingency funds as well as Air Force and Navy R&D funds, a formal Memorandum of Understanding (MoU) was drafted and sent around for signatures beginning in early November 1954. The MoU was originated by Trevor Gardner (Air Force Special Assistant for Research and Development), and was forwarded, respectively, for the signatures of J. H. Smith Jr.[29] (Assistant Secretary of the Navy [Air]) and Hugh L. Dryden (Director of the NACA). Dryden signed the MoU on 23 December 1954, and returned executed copies to the Air Force and Navy.[30]

The MoU (attached as an appendix to this monograph) provided that technical direction of the research project would be the responsibility of the NACA, acting "… with the advice and assistance of a Research Airplane Committee" composed of one representative each from the Air Force, Navy, and the NACA. Administration of the design and construction phases of the project was assigned to the Air Force. The NACA would conduct the flight research, with extensive support from the Air Force Flight Test Center. The Navy was essentially left paying 25 percent of the bills with little active roll in the project, although it would later supply biomedical expertise and a single pilot. The NACA and the Research Airplane Committee were charged with the responsibility for disseminating the research results to the military services and aircraft industry as appropriate based on various security aspects. The concluding statement on the MoU was: "Accomplishment of this project is a matter of national urgency."[31]

It should be noted that it was not unusual in the late 1940s and early 1950s for the military services to fund the development and construction of aircraft for the NACA to use in its flight test programs. This was how most of the testing on the X-1 and others had been accomplished. The eventual X-15 would be the fastest, highest-flying, and most expensive of these joint projects.[32]

After the signed copies of the MoU were returned to all participants, the Department of Defense authorized the Air Force to issue invitations to contractors having experience in the development of fighter-type aircraft to participate in the design competition. After the Christmas holidays, on 30 December, the Air Force sent invitation-to-bid letters to 12 prospective contractors; Bell, Boeing, Chance-Vought, Consolidated (Convair), Douglas, Grumman, Lockheed, Martin, McDonnell, North American, Northrop, and Republic. The letter asked those interested in bidding to notify Wright Field by 10 January 1955, and to attend a bidder's conference on 18 January 1955.[33]

Attached to the letter were a preliminary outline specification, an abstract of the Langley preliminary study, a discussion of possible engines, a list of data requirements, and a cost outline statement. Each bidder was required to satisfy various requirements set forth, except in the case of the NACA abstract which was presented as "… representative of possible solutions."[34]

Grumman, Lockheed, and Martin expressed little interest in the competition and did not attend the bidder's conference, leaving nine possible competitors. At the bidders' conference, representatives from the contractors met with NACA and Air Force personnel to discuss the competition and the basic design requirements.

During the bidders' conference, the airframe manufacturers were informed that one prime proposal and one alternate proposal (that might offer an unconventional but superior solution to the problems involved) would be accepted from each company. It also was noted that an engineering study, only, would be required for a modified aircraft where an observer could be substituted for the research instrumentation (a Navy requirement); that a weight allowance of 800 pounds, a volume of 40 cubic feet, and a power requirement of 2.25 kilowatts (kW) needed to be provided for research instrumentation; and that the winning design would have to be built in 30 months and be capable of attaining speeds of Mach 6 and altitudes of 250,000 feet. Following the preliminary statements concerning the bidding, NACA personnel briefed the various companies in attendance on new information that had resulted from late 1954 wind tunnel research that had taken place at Langley.

Subsequently, between the bidders' conference and the 9 May submission deadline, Boeing, Chance-Vought, Convair, Grumman, McDonnell, and Northrop notified the Air Force that they did not intend to submit formal proposals. This left Bell, Douglas, North American, and Republic. During this period, representatives from these companies met with NACA personnel on numerous occasions and reviewed technical information on various aspects of the forthcoming research airplane. The NACA also provided these contractors with further information gained as a result of wind tunnel tests in the Ames 10-by-14 inch supersonic tunnel and the Langley Mach 4 blowdown tunnel.

On 17 January 1955, NACA representatives met with Air Force personnel at Wright Field and were informed that the research airplane was identified as Air Force Project 1226 and would be officially designated X-15.

The Power Plant Laboratory had originally listed the Aerojet XLR73, Bell XLR81, North American NA-5400 (an engine in early development, still lacking a military designation), and the Reaction Motors XLR10 (and its variants, including the XLR30) as engines that the airframe competitors could use in their designs. Early in January, the laboratory had become concerned that the builders of engines other than those listed might protest the exclusion of their products. Consequently there emerged an explanation and justification of the engine selection process. It appeared that the engineers had confidence in the ability of the XLR81 and XLR73 to meet airplane requirements, had doubts about the suitability of the XLR25 (a Curtiss-Wright product), and held the thrust potential of the XLR8 and XLR11 (similar engines) in low repute. For practical purposes, this exhausted the available Air Force-developed engines suitable for manned aircraft. The XLR10 and NA-5400 were the only Navy-developed engines viewed as acceptable in terms of the competition.[35]

Earlier, the engine manufacturers had been contacted for specific information about the

engines originally listed as suitable for the X-15 program,[36] and this information was distributed to all four prospective airframe contractors.[37] Due to its early development status, there was little data available for the North American NA-5400, and the Reaction Motors XLR10 was "not recommended" at the suggestion of the engine manufacturer itself. On 4 February each of the prospective engine contractors (Aerojet, Bell, North American, and Reaction Motors) was asked to submit an engine development proposal.[38] Based on this, the Air Force very slightly relaxed the rigid limitations on engine selection, instructing competitors that "… if … an engine not on the approved list offers sufficient advantage, the airframe company may, together with the engine manufacturer, present justification for approval …" to the Air Force.[39]

On 9 May 1955, Bell, Douglas, North American, and Republic submitted their proposals to the Air Force. Two days later the technical data was distributed to the evaluation groups with a request that results be returned by 22 June.[40] The final evaluation meeting was scheduled for 25 July at Wright Field.[41]

Shortly thereafter, Hartley A. Soulé, as Chairman of the NACA evaluation group, sent the evaluation rules and processes to the NACA laboratories. The evaluation would be based on the technical and manufacturing competency of each contractor, schedule and cost estimates, design approach, and the research utility of each design. In order to expedite the evaluation, each of the NACA laboratories was assigned specific items to consider with responses to be returned to Soulé no later than 13 June.

The evaluation of the engine would be made at the same time, but would be conducted separate from that of the airframe contractor, with the possibility that the chosen engine might not be the one selected by the winning airframe contractor.

On 10 June the HSFS results were sent to Soulé, based on the design approach and research utility aspects of the airframe, flight control system, propulsion unit, crew provisions, handling and launching, and miscellaneous systems. The proposals were ranked: (1) Douglas; (2) North American; (3) Bell; and (4) Republic. The proposals from Douglas and North American were considered almost equal on the basis of points.

The Ames final evaluation, on 13 June 1955, ranked the proposals: (1) North American; (2) Douglas; (3) Bell; and (4) Republic. The North American structure was considered to be more representative of future aircraft and thus superior in terms of research utility. Douglas retained a simple and conventional magnesium structure, but in so doing avoided the very thermodynamic problems the research effort wished to explore.

The 14 June final evaluation from Langley ranked the proposals: (1) North American; (2) Douglas; (3) Republic; and (4) Bell. Langley felt that while the magnesium wing structure of Douglas was feasible, it was feared that local hot spots caused by irregular aerodynamic heating could weaken the structure and be subject to failure. North American's use of Inconel X was believed to be an advantage.

The final order representing the overall NACA evaluation was (1) North American; (2) Douglas; (3) Bell; and (4) Republic. All of the laboratories involved in this portion of the evaluation considered both the North American and Douglas proposals to be much superior to those submitted by Bell and Republic.

As with the NACA evaluations, the Air Force found little difference between the Douglas and North American designs, point-wise, with both proposals significantly superior to those of Bell and Republic. The Navy evaluation found much the same thing, ranking the proposals: (1) Douglas; (2) North American; (3) Republic; and (4) Bell.

On 26-28 July, the Air Force, Navy, and NACA evaluation teams met to coordinate their separate results. The Air Force and the NACA concluded that the North American proposal best accommodated their requirements. Accordingly, the Navy decided not to be put in the position of casting the dissenting vote and after short deliberation, agreed to go along with the decision of the Air Force and the NACA. A combined meeting of the Air Force, Navy, and the NACA was held at NACA Headquarters on 12 August for the final briefing on the evaluation. Later, the Research Airplane Committee met, accepted the findings of the evaluation groups, and agreed to present the recommendation to the Department of Defense.

Interestingly, the North American proposal was by far the most expensive. The estimated costs for three aircraft plus one static test article and supporting equipment were: Bell, $36.3 million; Douglas, $36.4 million; Republic, $47 million; and North American, $56.1 million.

Because the estimated costs submitted by North American were far above the amount allocated for the project, the Research Airplane Committee included a recommendation for a funding increase that would need to be approved before the actual contract was signed. A further recommendation, one that would later take on greater importance, called for relaxing the proposed schedule by up to one-and-one-half years. These recommendations were sent to the Assistant Secretary of Defense for Research and Development.

Events took an unexpected twist on 23 August when the North American representative in Dayton verbally informed the Air Force that the company wished to withdraw its proposal. On 30 August, North American sent a letter to the Air Force formally requesting that the company be allowed to withdraw from consideration.[42]

The Vice President and Chief Engineer for North American, Raymond H. Rice, wrote to the Air Force on 23 September and explained that the company had decided to withdraw from the competition because it had recently won new bomber and long range interceptor competitions and also had increased activity relating to its on-going F-107 fighter. Having undertaken these projects, North American said it would be unable to accommodate the fast engineering man-hours build-up that would be required to support the desired schedule. Rice went on that, "… due to the apparent interest that has subsequently been expressed in the North American design, the contractor [North American] wishes to extend two alternate courses which have been previously discussed with Air Force personnel: The engineering man-power work load schedule has been reviewed and the contractor wishes to point out that Project 1226 could be handled if it were permissible to extend the schedule… over an additional eight month period; in the event the above time extension is not acceptable and in the best interest of the project, the contractor is willing to release the proposal data to the Air Force at no cost."[43]

As it turned out, the possibility of extending the schedule had already been approved on 12 August, allowing North American to withdraw its previous letter of retraction once it had been officially informed that it had won the contract.[44] Accordingly, on 30 September 1955, the Air Force formally notified North American that its design had been selected as the winner. The other bidders were consequently notified of North American's selection and thanked for their participation.[45]

By 11 October, the estimate from North American had been reduced from $56,000,000 to $45,000,000 and the maximum annual funds requirement from $26,000,000 to $15,000,000. Shortly thereafter, the Department of Defense released the funds needed to start work. More meetings between the Air Force, the NACA, and North American were held on 27-28 October, largely to define changes to the aircraft configuration. On 18 November, letter contract AF33(600)-31693 was sent to North American, and an executed copy was returned on 8 December 1955.[46] The detailed design

and development of the hypersonic research airplane had been underway for just under a year at this point.[47]

On 1 December 1955, a series of actions[48] began that resulted in letter contract AF33(600)-32248 being sent to Reaction Motors, effective on 14 February 1956. Its initial allocation of funds totaled $3,000,000, with an eventual expenditure of about $6,000,000 foreseen as necessary for the delivery of the first flight engine.[49]

A definitive contract for North American was completed on 11 June 1956, superseding the letter contract and two intervening amendments. At that time, $5,315,000 had been committed to North American. The definitive contract allowed the eventual expenditure of $40,263,709 plus a fee of $2,617,075. For this sum, the government was to receive three X-15 research aircraft, a high speed and a low speed wind tunnel model program, a free-spin model, a full-size mockup, propulsion system tests and stands, flight tests, modification of a B-36 carrier aircraft, a flight handbook, a maintenance handbook, technical data, periodic reports of several types, ground handling dollies, spare parts, and ground support equipment. Exclusive of contract costs were fuel and oil, special test site facilities, and expenses to operate the B-36. The delivery date for the X-15s was to be 31 October 1958. The quantity of aircraft had been determined by experience; it had been noted during earlier research aircraft programs that two aircraft were enough to handle the anticipated workload, but three assured that the test pace could be maintained even with one aircraft down.[50] This lesson has been largely forgotten in our current budget-conscious era.

A final contract for the engine, the prime unit of government furnished equipment, was effective on 7 September 1956. Superseding the letter contract of February, it covered the expenditure of $10,160,030 plus a fee of $614,000.[51] For this sum, Reaction Motors agreed to deliver one engine, a mockup, reports, drawings, and tools.

Chapter 1
Notes and
References

[1] On 3 March 1915, Congress passed a Public law establishing "an Advisory Committee for Aeronautics." As stipulated in the Act, the purpose of this committee was "… to supervise and direct the scientific study of the problems of flight with a view to their practical solution."

[2] John V. Becker, "The X-15 Program in Retrospect" (paper presented at the 3rd Eugen Sänger Memorial Lecture, Bonn, Germany, 4-5 December 1968), p. 1.

[3] This ignores the research by the Air Force and Bell Aircraft into such skip-glide concepts as Bomber-Missile (BoMi) and Rocket-Bomber (RoBo), both of which were highly classified, and ultimately would be folded into the Dyna-Soar program.

[4] The accepted standard at the time was to report altitudes in statute miles.

[5] This had been the Langley Memorial Aeronautical Laboratory until July 1948 when the "Memorial" was dropped.

[6] John V. Becker, "The X-15 Program in Retrospect" (a paper presented at the 3rd Eugen Sänger Memorial Lecture, Bonn, Germany, 4-5 December 1968), p. 2.

[7] Letter from John V. Becker to Dennis R. Jenkins, 12 June 1999.

[8] Preliminary Outline Specification for High-Altitude, High-Speed Research Airplane, NACA Langley, 15 October 1954.

[9] General Requirements for a New Research Airplane, NACA Langley, 11 October 1954.

[10] John V. Becker, "The X-15 Program in Retrospect" (paper presented at the 3rd Eugen Sänger Memorial Lecture, Bonn, Germany, 4-5 December 1968), p. 2.

[11] This was well below the B-36's maximum useful load of 84,000 pounds, especially when it is considered that all military equipment on the B-36 would have been removed as well. But the conservative estimate allowed the carrier aircraft to fly high and fast enough to provide a decent launch platform for the research airplane.

[12] Dr. Hugh L. Dryden, "General Background of the X-15 Research Airplane Project" (a paper presented at the NACA Conference on the Progress of the X-15 Project, Langley Aeronautical Laboratory, Langley Field, Virginia, 25-26 October 1956), pp. xvii-xix and 3-9.

[13] John V. Becker, "Review of the Technology Relating to the X-15 Project" (a paper presented at the NACA Conference on the Progress of the X-15 Project, Langley Aeronautical Laboratory, Langley Field, Virginia, 25-26 October 1956, p. 3; John V. Becker, "The X-15 Program in Retrospect" (paper presented at the 3rd Eugen Sänger Memorial Lecture, Bonn, Germany, 4-5 December 1968), p. 7.

[14] John V. Becker, "The X-15 Program in Retrospect" (paper presented at the 3rd Eugen Sänger Memorial Lecture, Bonn, Germany, 4-5 December 1968), p. 2.

[15] Dr. Hugh L. Dryden, "Toward the New Horizons of Tomorrow," First von Kármán Lecture, Astronautics, January 1963.

[16] John V. Becker, "Review of the Technology Relating to the X-15 Project" (a paper presented at the NACA Conference on the Progress of the X-15 Project, Langley Aeronautical Laboratory, Langley Field, Virginia, 25-26 October 1956, pp. 4-5.

[17] These same trade studies would be repeated many times during the concept definition for Space Shuttle.

[18] Inconel X® is a nickel-chromium alloy whose name is a registered trademark of Huntington Alloy Products Division, International Nickel Company, Huntington, West Virginia.

[19] Dr. Hugh L. Dryden, "General Background of the X-15 Research Airplane Project" (a paper presented at the NACA Conference on the Progress of the X-15 Project, Langley Aeronautical Laboratory, Langley Field, Virginia, 25-26 October 1956), pp. xvii-xix.

[20] Memorandum from J. W. Rogers, Liquid Propellant and Rocket Branch, Rocket Propulsion Division, Air Research and Development Command (ARDC), to Lieutenant Colonel L. B. Zambon, Power Plant Laboratory, Wright Air Development Center (WADC), 13 July 1954, in the files of the AFMC History Office, Wright-Patterson AFB, Ohio.

[21] Letter from Colonel P. F. Nay, Acting Chief, Aeronautics and Propulsion Division, Deputy Commander of Technical Operations, ARDC, to Commander, WADC, 29 July 1954, subject: New Research Aircraft, in the files of the AFMC History Office, Wright-Patterson AFB, Ohio.

[22] Memorandum from J. W. Rogers, Liquid Propellant and Rocket Branch, Rocket Propulsion Division, Power Plant Laboratory, to Chief, Non-Rotating Engine Branch, Power Plant Laboratory, WADC, 11 August 1954, subject: Conferences on 9 and 10 August 1954 on NACA Research Aircraft-Propulsion System, in the files of the AFMC History Office, Wright-Patterson AFB, Ohio.

[23] A published summary of the 9 July NACA presentations did not appear until 14 August.

[24] Letter from Colonel V. R. Haugen to Commander, ARDC, 13 August 1954, in the files of the AFMC History Office, Wright-Patterson AFB, Ohio.

[25] Memorandum from R. L. Schulz, Technical Director of Aircraft, to Chief, Fighter Aircraft Division, Director of WSO, WADC, not dated (presumed about 13 August 1954), in the files of the AFMC History Office, Wright-Patterson AFB, Ohio.

[26] Letter from Brigadier General Benjamin S. Kelsey, Deputy Director of R&D, DCS/D, USAF, to Commander, ARDC, 4 October 1954, subject: New Research Aircraft, in the files of the AFMC History Office, Wright-Patterson AFB, Ohio.

[27] Letter from A. Scott Crossfield to Dennis R. Jenkins, 30 June 1999.

[28] Dr. Hugh L. Dryden, "General Background of the X-15 Research Airplane Project" (a paper presented at the NACA Conference on the Progress of the X-15 Project, Langley Aeronautical Laboratory, Langley Field, Virginia, 25-26 October 1956), pp. xvii-xix.

[29] It was common practice in the 1950s to only record the last name and initials for individuals on official correspondence. Where possible first names are provided; but in many cases a first name cannot be definitively determined from available documentation.

[30] Memorandum from Trevor Gardner, Special Assistant for R&D, USAF, to J. H. Smith Jr., Assistant Secretary Navy

(Air), 9 November 1954, subject: Principles for the Conduct of a Joint Project for a New High Speed Research Airplane; Letter from J. H. Smith Jr., Assistant Secretary Navy (Air), to Dr. Hugh L. Dryden, Director of NACA, 21 December 1954; Letter from Dr. Hugh L. Dryden to Trevor Gardner returning a signed copy of the MoU, 23 December 1954, in the files of the NASA History Office, Washington, DC.

[31] Memorandum of Understanding, signed by Hugh L. Dryden, Director of NACA, J. H. Smith Jr., Assistant Secretary Navy (Air), and Trevor Gardner, Special Assistant for R&D, USAF, 23 December 1954, subject: Principles for the Conduct by the NACA, Navy, and Air Force of a Joint Project for a New High-Speed Research Airplane, in the files of the NASA History Office, Washington, DC.

[32] Walter C. Williams, "X-15 Concept Evolution" (a paper in the *Proceedings of the X-15 30th Anniversary Celebration*, Dryden Flight Research Facility, Edwards, California, 8 June 1989, NASA CP-3105), p. 11.

[33] Memorandum from A. L. Sea, Assistant Chief, Fighter Aircraft Division, to Director of Weapons Systems Office, WADC, 29 December 1954, subject: New Research Aircraft, in the files of the AFMC History Office, Wright-Patterson AFB, Ohio.

[34] Letter from Colonel C. F. Damberg, Chief, Aircraft Division, Air Materiel Command, to Bell Aircraft Corporation, et al., 30 December 1954, subject: Competition for New Research Aircraft; Memorandum from A. L. Sea, to Director of WSO, WADC, 29 December 1954; Letter from J. B. Trenholm, Chief, New Development Office, Fighter Aircraft Division, Director of WSO, WADC, to Commander, ARDC, 13 January 1955, subject: New Research Aircraft, in the files of the AFMC History Office, Wright-Patterson AFB, Ohio.

[35] Memorandum from J. W. Rogers to Chief, Power Plant Laboratory, 4 January 1955, in the files of the AFMC History Office, Wright-Patterson AFB, Ohio.

[36] Letter from W. P. Turner, Manager, Customer Relations and Contracts Division, Reaction Motors, Inc. (hereafter cited as RMI), to Commander, AMC, 3 February 1955, subject: XLR30 Rocket Engine—Information Concerning, in the files of the AFMC History Office, Wright-Patterson AFB, Ohio.

[37] Letter from J. B. Trenholm, Chief, New Development Office, Fighter Aircraft Division, Director of WSO, WADC, to Bell, et al., 22 March 1955, subject: Transmittal of Data, Project 1226 Competition, in the files of the AFMC History Office, Wright-Patterson AFB, Ohio.

[38] Letter from R. W. Walker, Chief, Power Plant Development Section, Power Plant Branch, Aero-Equipment Division, AMC, to RMI, 4 February 1955, subject: Power Plant for New Research Airplane, in the files of the AFMC History Office, Wright-Patterson AFB, Ohio.

[39] Letter from Colonel C. F. Damberg, Chief, Aircraft Division, AMC, to Bell, et al., 2 February 1955, subject: Project 1226 Competition, in the files of the AFMC History Office, Wright-Patterson AFB, Ohio.

[40] Letter from Hugh L. Dryden, Director of NACA, to Deputy Director of R&D, DCS/D, USAF, 20 May 1955, no subject; Letter from Rear Adm. R. S. Hatcher, Assistant Chief of R&D, BuAer, USN, to Commander, WADC, 31 May 1955, subject: Agreements Reached by "Research Airplane Committee," on Evaluation Procedure for X-15 Research Airplane Proposals, in the files of the AFMC History Office, Wright-Patterson AFB, Ohio.

[41] Memorandum from Brigadier General Howell M. Estes Jr., Director of WSO, to Director of Laboratories, WADC 28 June 1955, subject: X-15 Evaluation, in the files of the AFMC History Office, Wright-Patterson AFB, Ohio.

[42] Memorandum from Colonel C. G. Allen, Chief, Fighter Aircraft Division, Director of Systems Management, ARDC, to Commander, WADC 23 August 1955, subject: X-15 Evaluation, in the files of the AFMC History Office, Wright-Patterson AFB, Ohio.

[43] Letter from R. H. Rice, Vice President and Chief Engineer, North American Aviation, Inc., to Commander, ARDC, 23 September 1955, subject: Project 1226, Research Airplane, in the files of the AFMC History Office, Wright-Patterson AFB, Ohio.

[44] X-15 WSPO Weekly Activity Report, 22 September 1955, in the files of the AFMC History Office, Wright-Patterson AFB, Ohio.

[45] Letter from Colonel C. F. Damberg, Chief, Aircraft Division, AMC, to North American Aviation, 30 September 1955, subject: X-15 Competition, in the files of the AFMC History Office, Wright-Patterson AFB, Ohio.

[46] X-15 WSPO Weekly Activity Report, 13 October 1955; 20 October 1955; 27 October 1955; and 15 December 1955; Letter from N. Shropshire, Director of Contract Administration, NAA, to Commander, AMC, 8 December 1955, subject: Letter Contract AF33(600)-31693, in the files of the AFMC History Office, Wright-Patterson AFB, Ohio.

[47] Dr. Hugh L. Dryden, "General Background of the X-15 Research Airplane Project" (a paper presented at the NACA Conference on the Progress of the X-15 Project, Langley Aeronautical Laboratory, Langley Field, Virginia, 25-26 October 1956), pp. xviii.

[48] Memorandum from Captain Chester E. McCollough Jr., Project Officer, New Development Office, Fighter Aircraft Division, Director of Systems Management, ARDC, to Chief, Non-Rotating Engine Branch, Power Plant Laboratory, Director of Laboratories, WADC, 1 December 1955, subject: Engine for X-15, in the files of the AFMC History Office, Wright-Patterson AFB, Ohio.

[49] Contract AF33(600)-32248, 14 February 1956, in the files of the AFMC History Office, Wright-Patterson AFB, Ohio.

[50] Contract AF33(600)-31693, 11 June 1956, in the files of the AFMC History Office, Wright-Patterson AFB, Ohio.

[51] The final fee paid to Reaction Motors was greater than the original estimate for the total engine development program; the definitive contract exceeded more than 20 times the original estimate, and more than twice the original program approval estimate. As events later demonstrated, even this erred badly on the side of underestimation.

Chapter 2

X-15 Design and Development

Harrison A. "Stormy" Storms, Jr. led the North American X-15 design team, along with project engineer Charles H. Feltz. These two had a difficult job ahead of them, for although giving the appearance of having a rather simple configuration, the X-15 was perhaps the most technologically complex single-seat aircraft of its day. Directly assisting Storms and Feltz was test pilot A. Scott Crossfield, who had worked for the NACA prior to joining North American with the intended purpose of working on the X-15 program. Crossfield describes Storms as "… a man of wonderful imagination, technical depth, and courage … with a love affair with the X-15. He was a tremendous ally and kept the objectivity of the program intact …." According to Crossfield, Feltz was "… a remarkable 'can do and did' engineer who was very much a source of the X-15 success story."[1]

Storms himself remembers his first verbal instructions from Hartley Soulé: "You have a little airplane and a big engine with a large thrust margin. We want to go to 250,000 feet altitude and Mach 6. We want to study aerodynamic heating. We do not want to worry about aerodynamic stability and control, or the airplane breaking up. So if you make any errors, make them on the strong side. You should have enough thrust to do the job." Adds Storms, "and so we did."[2]

Crossfield's X-15 input proved particularly noteworthy during the early days of the development program as his experience permitted the generation of logical arguments that led to major improvements to the X-15. He played a key role, for instance, in convincing the Air Force that an encapsulated ejection system was both impractical and

By the time of the first industry conference in 1956, this was the design baseline for the North American X-15. Note the tall vertical stabilizer, and the fact that it does not have the distinctive wedge shape of the final unit. Also notice how far forward the fuselage tunnels extend—well past the canopy. (NASA)

SYS-447L **NORTH AMERICAN X-15 RESEARCH AIRPLANE**

unnecessary. His arguments in favor of an ejection seat, capable of permitting safe emergency egress at speeds between 80 mph and Mach 4 and altitudes from sea level to 120,000 feet, saved significant money, weight, and development time.

There has been considerable interest over whether Crossfield made the right decision in leaving the NACA since it effectively locked him out of the high-speed, high-altitude portion of the X-15 flight program. Crossfield has no regrets: "… I made the right decision to go to North American. I am an engineer, aerodynamicist, and designer by training … While I would very much have liked to participate in the flight research program, I am pretty well convinced that I was needed to supply a lot of the impetus that allowed the program to succeed in timeliness, in resources, and in technical return. … I was on the program for nine years from conception to closing the circle in flight test. Every step: concept, criteria, requirements, performance specifications, detailed specifications, manufacturing, quality control, and flight operations had all become an [obsession] to fight for, protect, and share—almost with a passion."[3]

Although the first, and perhaps the most influential pilot to contribute to the X-15 program, Crossfield was not the only one to do so. In fact, all of the initially assigned X-15 pilots participated in the development phases, being called on to evaluate various operational systems and approaches, as well as such factors as cockpit layout, control systems, and guidance schemes. They worked jointly with engineers in conducting the simulator programs designed to study the aspects of planned flight missions believed to present potential difficulties. A fixed-base simulator was developed at North American's Los Angeles facility, containing a working X-15 cockpit and control system that included actual hydraulic and control-system hardware. Following use at North American, it was subsequently relocated to the Flight Research Center[4] (FRC) at Edwards AFB. Once flight research began, the simulator was constantly refined with the results of the flight test program, and late in its life the original analog computers were replaced by much faster digital units. For the life of the program, every X-15 flight was preceded by 10-20 hours in the simulator.

A ground simulation of the dynamic envi-

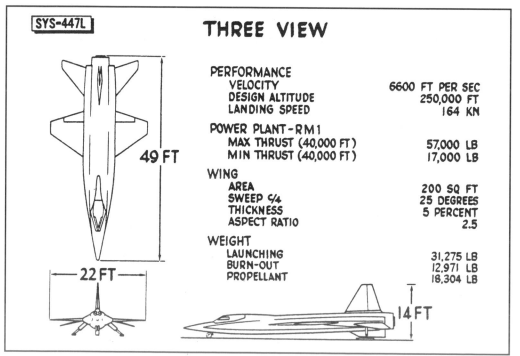

SYS-447L

THREE VIEW

PERFORMANCE
VELOCITY   6600 FT PER SEC
DESIGN ALTITUDE   250,000 FT
LANDING SPEED   164 KN

POWER PLANT-RM1
MAX THRUST (40,000 FT)   57,000 LB
MIN THRUST (40,000 FT)   17,000 LB

WING
AREA   200 SQ FT
SWEEP c/4   25 DEGREES
THICKNESS   5 PERCENT
ASPECT RATIO   2.5

WEIGHT
LAUNCHING   31,275 LB
BURN-OUT   12,971 LB
PROPELLANT   18,304 LB

49 FT

22 FT

14 FT

One of the more controversial features of the North American design was the fuselage tunnels that carried the propellant lines and engine controls around the full monocoque propellant tanks, shown in this 1956 sketch. Originally these tunnels extended forward ahead of the cockpit, and the NACA worried they would create unacceptable vortices. (NASA)

ronment was provided by use of the Navy centrifuge at the Naval Air Development Center (NADC) Johnsville, Pennsylvania. Over 400 simulated reentries[5] were flown during an initial round of tests completed on 12 July 1958; Iven Kincheloe, Joe Walker, Scott Crossfield, Al White, Robert White, Neil Armstrong, and Jack McKay participated. The primary objective of the program was to assess the pilot's ability to make emergency reentries under high dynamic conditions following a failure of the stability augmentation system. The results were generally encouraging.[6]

When the contracts with North American had been signed, the X-15 was some three years away from actual flight test. Although most of the basic research into materials and structural science had been completed, a great deal of work remained to be accomplished. This included the development of fabrication and assembly techniques for Inconel X and the new hot-structure design. North American met the challenge of each problem with a practical solution, and eventually some 2,000,000 engineering man-hours and 4,000 wind tunnel hours in 13 different wind tunnels were logged.

The original North American proposal gave rise to several questions which prompted a meeting at Wright-Patterson AFB on 24-25 October 1955. Subsequent meetings were held at the North American Inglewood plant on 28-29 October and 14-15 November. Major discussion items included North American's use of fuselage tunnels and all-moving horizontal stabilizers (the "rolling-tail"). The rolling-tail operated differentially to provide roll control, and symmetrically to provide pitch control; this allowed the elimination of conventional ailerons. North American had gained considerable experience with all-moving control surfaces on the YF-107A fighter. In this instance the use of differentially operated surfaces simplified the construction of the wing, and allowed elimination of protuberances that would have been necessary if aileron actuators had been incorporated in the thin wing. Such protuberances would have disturbed the airflow and created another heating problem.

One other significant difference between the configuration of the NACA design and that of the actual X-15 stemmed from North American's use of full-monocoque propellant tanks in the center fuselage and the use

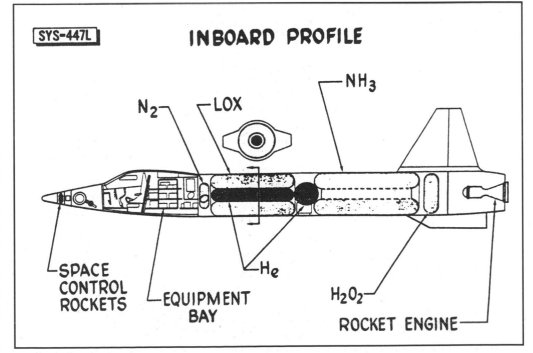

The interior layout of the fuselage did not change much after the 1956 conference. Note the helium tank located in the middle of the LOX tank. The hydrogen-peroxide (H2O2) was used to power the turbopump on the XLR99 rocket engine. (NASA)

of tunnels on both sides of the fuselage to accommodate the propellant lines and engine controls that ordinarily would have been contained within the fuselage. The NACA expressed concern that the tunnels might create undesirable vortices that would interfere with the vertical stabilizer, and suggested that the tunnels be kept as short as possible in the area ahead of the wing. North American agreed to make the investigation of the tunnels' effects a subject of an early wind tunnel-model testing program.[7]

During the spring and summer of 1956, several scale models were exposed to rather intensive wind tunnel tests. A 1/50-scale-model was tested in the 11-inch hypersonic and 9-inch blowdown tunnels at Langley, and another in a North American wind tunnel. A 1/15-scale model was also tested at Langley and a rotary-derivative model was tested at Ames. The various wind tunnel programs included investigations of the speed

brakes, horizontal stabilizers without dihedral, several possible locations for the horizontal stabilizer, modifications of the vertical stabilizer, the fuselage tunnels, and control effectiveness, particularly of the rolling-tail. Another subject in which there was considerable interest was determining the cross-section radii for the leading edges of the various surfaces.

On 11 June 1956, North American received a production go-ahead for the three X-15 airframes (although the first metal was not cut for the first aircraft until September). Four days later, on 15 June 1956, the Air Force assigned three serial numbers (56-6670 through 56-6672) to the X-15 program.[8]

By July, the NACA felt that sufficient progress had been made on the X-15 development to make an industry conference on the project worthwhile.[9] The first Conference on the Progress of the X-15 Project was held

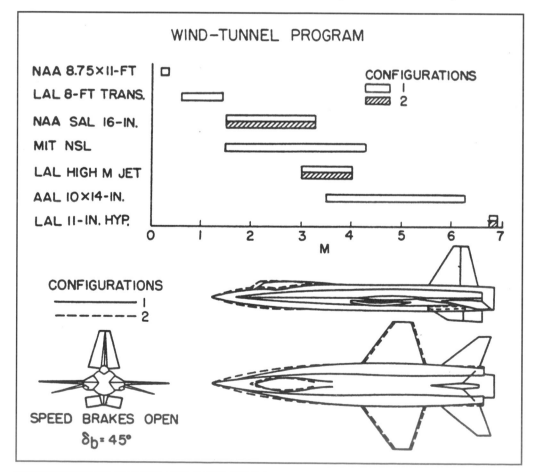

Seven different wind tunnels are represented in this chart showing how the extreme front of the fuselage tunnels began to be modified. Note the large speed brakes on the vertical stabilizer.

"LAL" on the chart is the Langley Aeronautical Laboratory, while "AAL" is the Ames Aeronautical Laboratory. (NASA)

at Langley on 25-26 October 1956. There were 313 attendees representing the Air Force, the NACA, Navy, various universities and colleges, and most of the major aerospace contractors. It was evident from the papers that a considerable amount of progress had already been made, but that a few significant problems still lay ahead.[10]

A comparison of the suggested configuration contained in the original NACA proposal and the North American configuration presented to the industry conference revealed that the span of the X-15 had been reduced from 27.4 feet to only 22 feet and that the North American fuselage had grown from the suggested 47.5-foot overall-length to 49 feet. North American followed the NACA suggestion by selecting Inconel X as the major structural material and in the design of a multispar wing with extensive use of corrugated webs.[11]

One of the papers summarized the aerodynamic characteristics that had been obtained by tests in eight different wind tunnels.[12] These tests had been made at Mach numbers ranging from less than 0.1 to about 6.9, and investigated such problems as the effects of speed brake deflection on drag, the lift-drag relationship of the entire aircraft, of individual components such as the wings and fuselage tunnels, and of combinations of individual components. One of the interesting products was a finding that almost half of the total lift at high Mach numbers would be derived from the fuselage tunnels. Another

result was the confirmation of the NACA's prediction that the original fuselage tunnels would cause longitudinal instability; for subsequent testing the tunnels had been shortened in the area ahead of the wing, greatly reducing the instability. Still other wind tunnel tests had been conducted in an effort to establish the effect of the vertical and horizontal tail surfaces on longitudinal, directional, and lateral stability.

It should be noted that wind tunnel testing in the late 1950s was, and still is, an inexact science. For example, small (3- to 4-inch) models of the X-15 were "flown" in the hypervelocity free-flight facility at Ames. The models were made out of cast aluminum, cast bronze, or various plastics, and were actually fairly fragile. Despite this, the goal was to shoot the model out of a gun at tremendous speeds in order to observe shock wave patterns across the shape. As often as not, what researchers saw were pieces of X-15 models flying down the range sideways. Fortunately, enough of the models remained intact to acquire meaningful data.[13]

Other papers presented at the industry conference dealt with research into the effect of the aircraft's aerodynamic characteristics on the pilot's control. Pilot-controlled simulation flights for the exit and reentry phases had been conducted; researchers reported that the pilots had found the early configurations nearly uncontrollable without damping, and that even with dampers the airplane possessed only minimum stability during parts

These charts show the expected temperatures and skin thickness for various parts of the X-15's fuselage. Note the large difference between top-side temperatures and those on the bottom of the fuselage. (NASA)

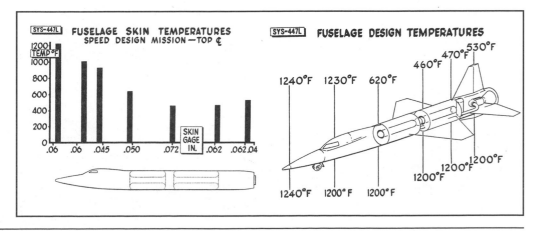

of the programmed flight plan. A program utilizing a free-flying model had proved low-speed stability and control to be adequate. Since some aerodynamicists had questioned North American's use of the rolling-tail instead of ailerons, the free-flying model had also been used to investigate that feature. The results indicated that the rolling-tail would provide the necessary lateral control.

Several papers presented at the conference dealt with aerodynamic heating. One of these was a summary of the experience gained with the Bell X-1B and X-2. The information was incomplete and not fully applicable to the X-15, but it did provide a basis for comparison with the results of the wind tunnel and analytical studies. Another paper dealt with the results of the structural temperature estimates that had been arrived at analytically. It was apparent from the contents of the papers that the engineers compiling them were confronted by a paradox—in order to attain an adequate and reasonably safe research vehicle, they had to foresee and compensate for the very aerodynamic heating problems that were to be explored by the completed aircraft.

In addition to the papers on the theoretical aspects of aerodynamic heating, a report was made on the structural design that had been accomplished at the time of the conference. Critical loads would be encountered during the accelerations at launch weight and during reentry into the atmosphere, but since maximum temperatures would be encountered only during the latter, the paper was largely confined to the results of the investigations of the load-temperature relationships that were anticipated for the reentry phase. The selection of Inconel X skin for the multispar box-beam wing was justified on the basis of the strength and favorable creep characteristics of that material at 1,200 degrees Fahrenheit. A milled bar of Inconel X was to be used for the leading edge since that portion of the wing acted as a heat sink. The internal structure of the wing was to be of titanium-alloy sheet and extrusion construction. The front and rear spars were to be flat web-channel sections with the intermediate spars and ribs of corrugated titanium webs.

For purposes of the tests the maximum temperature differences between the upper and lower wing surfaces had been estimated to be 400 degrees Fahrenheit and that between the skin and the center of the spar as 960 degrees Fahrenheit. Laboratory tests indicated that such differences could be tolerated without any adverse effects on the structure. Other tests had proven that thermal stresses for the Inconel-titanium structure were less than those encountered in similar structures constructed entirely on Inconel X. Full-scale tests had been made to determine the effects of temperature on the buckling and ultimate strength of a box beam. Simply heating the test structure produced no surface buckles. Compression buckles had appeared when ultimate loads were applied at normal temperatures but the buckles disappeared with the removal of the load. Tests at higher temperatures and involving large temperature

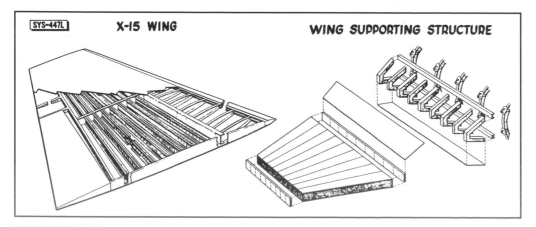

The wing of the X-15 was constructed from Inconel X skins over a titanium structure. Unlike many aircraft, there was not a continuous spar across both wings. Instead, each wing was bolted to the fuselage. (NASA)

differences had finally led to the failure of the test box, but it seemed safe to conclude that "… thermal stresses had very little effect on the ultimate strength of the box."

Tests similar to those conducted on the wing structure had also been performed on the horizontal stabilizer. The planned stabilizer structure differed from the wing in that it incorporated a stainless steel spar about halfway between the leading and trailing edges, and an Inconel X spar three and one-half inches from the leading edge. The remainder of the internal structure consisted of titanium components and the skin was Inconel X sheet. Tests of the stabilizer had indicated that a design which would prevent all skin buckling would be inordinately heavy, so engineers decided to tolerate temporary buckles. The proposed stabilizer had flutter characteristics that were within acceptable limits.

The front and rear fuselage were semimonocoque structures of titanium ribs, Inconel X outer skin, and an inner aluminum skin insulated with spun glass. The integral propellant tanks in the center fuselage were of full monocoque construction. The full monocoque design used only slightly thicker skins than the semimonocoque design, possessed adequate heat sink properties, and reduced stresses caused by temperature differences by placing all of the material at the surface. It seemed, therefore, that the resulting structure was ideal for use as a pressure tank. The thickness of the monocoque walls would also make sealing easier and leaks less likely.

The fuselage side tunnels presented yet another problem. As the tunnels would protect the side portions of the propellant tanks from aerodynamic heating, the sides would not expand as rapidly as the areas exposed to the air, and another undesirable compressive stress had to be anticipated. It was thought that beading the skin of the areas protected by the tunnels would provide a satisfactory solution, but beading introduced further complications by reducing the structure's ability to carry pressure loads. Ultimately, however, the techniques proved successful.

Like most rocket engines of the period, the XLR99 would use liquid oxygen as an oxidizer, and a non-cryogenic fuel, in this case anhydrous ammonia.[14] Each of the two main propellant tanks was to be divided into three compartments by curved bulkheads; the two compartments furthest from the aircraft center of gravity were equipped with slosh baffles. Plumbing was to be installed in a single compartment, the compartment sealed by a bulkhead, and the process repeated until all the compartments were completed. The tank ends were to be semicurved in shape to keep them as flat as possible, to reduce weight, and to permit thermal expansion of the tank shell. This entire structure was to be of welded Inconel X.

The expected acceleration of the X-15 presented several unique human factors concerns early in the program. It was expected that the pilot would be subjected to an acceleration of up to 5g. It seemed advisable to develop a

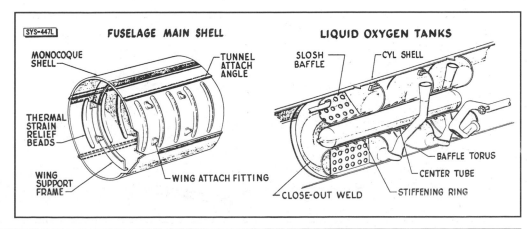

One of the innovations proposed by North American was the use of monocoque propellant tanks, leading to the use of the controversial fuselage tunnels. The forward-most part of the LOX tank was equipped with slosh baffles. (NASA)

side-stick controller that would allow the pilot's arm to be supported by an armrest while still allowing him of full control over the aircraft.[15] Coupled with the fact that there were two separate attitude-control systems on the X-15, this resulted in a unique control stick arrangement. A conventional center stick, similar to that installed in most fighter-type aircraft of the era, was connected to the aerodynamic control surfaces through a stability-augmentation (damper) system. A side-stick controller on the right console was connected to the same aerodynamic control surfaces and augmentation system. Either stick could be used interchangeably, although the flight manual[16] describes using the center stick "during normal periods of longitudinal and vertical acceleration." The center stick was occasionally omitted from flights later in the flight research program based on pilot preferences. Another side-stick controller on the left console operated the so-called "ballistic control" system[17] (thrusters) that provided attitude control at high altitudes. The flight manual warns that "velocity tends to sustain itself after the stick is returned to the neutral position. A subsequent stick movement opposite to the initial one is required to cancel the original attitude change."

At the time of the industry conference in 1956, the design for the X-15 side controller had not been definitely established but a summary of the previous experience with such controllers was available. Experimental controllers had been installed on a Grumman F9F-2, Lockheed TV-2, Convair F-102, and on a simulator. The pilots who had tried side controllers had reported no difficulty in maneuvering, but they generally felt that greater efforts would have to be made to eliminate backlash and to control friction forces; they had also urged that efforts be made to give the side controllers a more "natural" feel.

Another problem which had not been thoroughly explored at the time of the 1956 conference concerned the proposed reaction controls that would be necessary for the X-15 as dynamic pressures decreased to the point where the aerodynamic controls would no longer be effective. Analog computer and ground simulator studies were then under way in an effort to determine the best relationship between the control thrust and the pilot's movement of the control stick. Attempts were also being made to determine the amount of propellant that would be required for the reac-

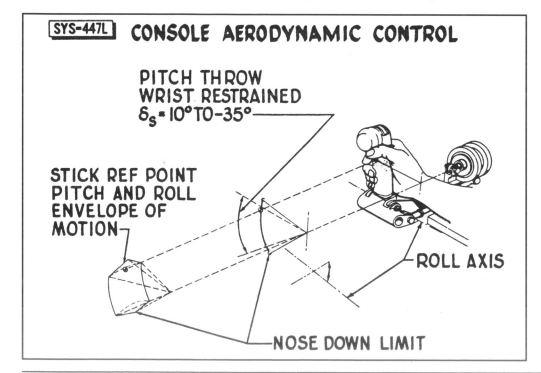

The X-15 contained two side-stick controllers; one for the aerodynamic controls (shown), and one on the other console for the reaction controls. Although the side-stick proved very successful on the X-15, it would be another 20 years before one was installed on an operational aircraft (the General Dynamics F-16). (NASA)

tion controls. No significant problems were uncovered during these early investigations, but it was clear that the pilot would have to give almost constant attention to such a control system and that pilots should be given extensive practice on simulators before being allowed to attempt actual flight.

Some of the anticipated difficulties in the field of instrumentation arose because available strain gauges were not considered satisfactory at the expected high temperatures and because of difficulties in recording the output of thermocouples. Large structural deformations of wings and empennage were to be recorded by cameras in special camera compartments. Another instrumentation problem arose because the sensing of static pressure, ordinarily difficult at high Mach numbers, was compounded in the case of the X-15 by heating that would be too great for any conventional probe and by the low pressure at the high altitudes to be explored. The answer was to develop a stable-platform-integrating-accelerometer system to provide velocity, altitude, pitch, yaw, and roll angle information.

Still another instrumentation difficulty was created by the desirability of presenting the

pilot with angle-of-attack and side slip information, especially for the critical exit and reentry periods. Any device to furnish this information would have to be located ahead of the aircraft's own flow disturbances, be structurally sound at elevated temperatures, accurate at low pressures, and cause a minimal flow disturbance so as not to interfere with the heat transfer studies that were to be conducted in the forward area of the fuselage. These requirements had resulted in the design of a ball-nose[18] capable of withstanding 1,200 degrees Fahrenheit. A six-inch diameter Inconel X sphere located in the extreme nose of the X-15 was gimbaled[19] and servo-driven in two planes. It had five openings: a total-head port opening directly forward and two pairs of angle-sensing ports in the pitch and yaw planes, located at an angle of 30 to 40 degrees from the central port. Pitch and yaw could be sensed as pressure differences and these differences were converted into signals that would cause the servos to realign the sphere in the relative wind.

Based largely on urgings from Scott Crossfield, the Air Force agreed to allow North American to design an ejection seat and to make a study justifying the selection

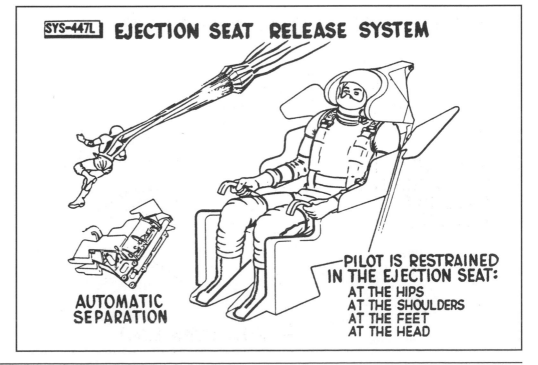

Although the ejection seat showed at the 1956 industry conference did not resemble the final unit used in the X-15s, the basic concepts remained the same. Restraining the pilot's head, arms, and legs during ejection at high dynamic pressures presented one of the major challenges to seat development. (NASA)

SYS-447L EJECTION SEAT RELEASE SYSTEM

AUTOMATIC SEPARATION

PILOT IS RESTRAINED IN THE EJECTION SEAT:
AT THE HIPS
AT THE SHOULDERS
AT THE FEET
AT THE HEAD

of a seat in preference to a capsule system.[20] Two main criteria had governed the selection of an escape system for the X-15, and these criteria were not necessarily complementary. The first requirement was that the system be the most suitable that could be designed while remaining compatible with the airplane. The second was that no system would be selected that would delay the development of the X-15 or leave the pilot without any method of escape when the time arrived for flight research. The four possible escape systems that were considered included cockpit capsules, nose capsules, a canopy shielded seat, and a stable-seat with a pressure-suit. An analysis of the expected flight hazards had indicated that because of the fuel exhaustion and low aerodynamic loads, the accident potential at peak speeds and altitudes was only about two percent of the total.

Capsule-like systems had been tried before, most notably in the X-2 where the entire forward fuselage could be detached from the rest of the aircraft. Model tests showed these to be very unstable and prone to tumble at a high rate of rotation. They also added a great deal of weight and complexity to the aircraft.[21]

The final decision for a stable-seat with a pressure-suit was made because most of the potential accidents could be expected to occur at speeds of Mach 4 or less, because system reliability always decreased with system complexity, and finally, because it was the system that imposed the smallest weight and size penalties upon the aircraft. The selected system would not function successfully at altitudes above 120,000 feet or speeds in excess of Mach 4, but designers, particularly Scott Crossfield, held that the aircraft itself would offer the best protection in the areas of the performance envelope where the seat-suit combination was inadequate.

Cockpit and instrument cooling, pressurization, suit ventilation, windshield defogging, and fire protection were all to be provided from a liquid nitrogen supply. Vaporization of the liquid nitrogen would keep the pilot's environment within comfortable limits at all times. An interesting aspect of the cooling problem was an estimate that only 1.5 percent of the system's capacity would be applied to the pilot; the remaining 98.5 percent was required for equipment. Cockpit temperatures were to be limited to no more than 150 degrees Fahrenheit, the maximum

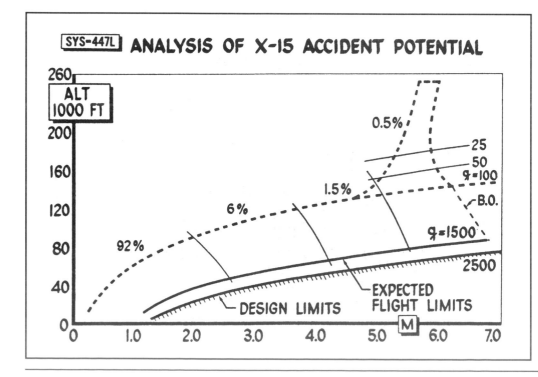

This chart shows that 92 percent of the expected X-15 accidents would happen below Mach 2 and 90,000 feet. This estimate supported Scott Crossfield's request to use an ejection seat and pressure suit instead of a more complex escape capsule. (NASA)

limit for some of the equipment. The pilot would not be subjected to that temperature, however, as the pressure suit ventilation would enable him to select a comfortable temperature level. Cockpit pressure was to be maintained at the 35,000 foot level.

The effects of flight accelerations upon the pilot's physiological condition and upon his ability to avoid inadvertent control movements had not been completely explored, but it was recognized that high accelerations could pose medical and restraint difficulties. In addition to the accelerations that would be encountered during the exit and reentry phases of the X-15's flights, a very high acceleration of short duration would be produced during the landings. This was a result of the location of the main skids at the rear of the aircraft. Once the skids touched down, the entire aircraft would act as if it were hinged at the skid attachment points and the nose section would slam downward. Reproduction of this landing acceleration on simulators showed that because of the short duration, no real problem existed. There were, however, numerous complaints about the severity of the jolts both in the simulator and once actual landings began.

The final paper presented to the 1956 industry conference was an excellent summary of the development effort and a review of the major problems that were known at that time. The author, Lawrence P. Greene from North American, considered flutter to be an unsolved problem, primarily because of a lack of basic data on aero-thermal-elastic relationships and because little experimental data was available on flutter at hypersonic Mach numbers. He pointed out that available data on high-speed flutter had been derived from experiments conducted at Mach 3 or less, and that not all of the data obtained at those speeds were applicable to the problems faced by the designers of the X-15. As it turned out, panel flutter was encountered early in the flight test program, leading to a change in the design criteria for high-speed aircraft. Another difficulty was the newness of Inconel X as a structural material and the necessity of experimenting with fabrication techniques that would permit its use as the primary structural material for the X-15. Problems were also expected to arise in connection with sealing materials, most of which were known to react unfavorably when subjected to high temperature conditions.[22] Although North American did encounter initial problems in using Inconel

Despite its performance potential, the basic cockpit design of the X-15 was quite conventional, with the exception of the side-stick controllers. The engine instrumentation on the lower left of the instrument panel would be different for the XLR11 flights. The addition of the MH-96 in the X-15-3 would necessitate some changes in the instrumentation. See page 63 for a photo. (NASA)

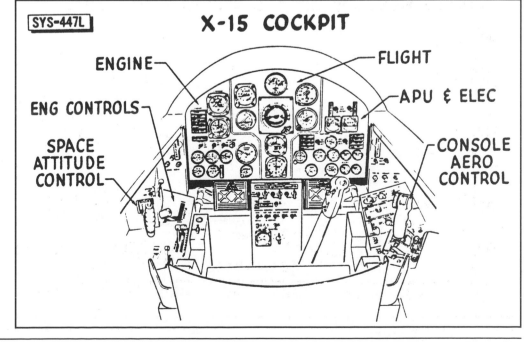

SYS-447L

X-15 COCKPIT

ENGINE

ENG CONTROLS

SPACE ATTITUDE CONTROL

FLIGHT

APU & ELEC

CONSOLE AERO CONTROL

X and titanium during the construction of the X-15, it was able to work through the difficulties with no major delays.

A development engineering inspection was held at the North American Inglewood plant on 12-13 December 1956. This inspection of a full-scale mockup was intended to reveal unsatisfactory design features before fabrication of the aircraft got under way. Thirty-four of the forty-nine individuals who participated in the inspection were representatives of the Air Force; twenty-two of them from WADC. The important role of the Air Force was also evident from the composition of the committee that would review the requests for alteration.[23] Major E. C. Freeman, of ARDC, served as committee chairman, Mr. F. Orazio of WADC and Lieutenant Colonel Keith G. Lindell of Air Force Headquarters were committee members, and Captain Chester E. McCollough, Jr. of the ARDC and Captain Iven C. Kincheloe, Jr. of the Air Force Flight Test Center (AFFTC) served as advisors. The Navy and the NACA each provided a single committee member; three additional advisors were drawn from the NACA.

The inspection committee considered 84 requests for alterations, rejected 12, and placed 22 in a category for further study. The majority of the 50 changes that were accepted were minor, such as the addition of longitudinal trim indications from the stick position and trim switches, relocation of the battery switch, removal of landing gear warning lights, rearrangement and redesign of warning lights, and improved markings for several instruments and controls.

Some of the most interesting comments were rejected by the committee. For instance, the suggestions that the aerodynamic and reaction controller motions be made similar, that the reaction controls be made operable by the same controller used for the aerodynamic controls, or that a third controller combining the functions of the aerodynamic and reaction controllers be added to the right console, were all rejected on the grounds that actual flight experience was needed with the controllers already selected before a decision could be made on worthwhile improvements or combinations. As two of the three suggestions on the controllers came from potential pilots of the X-15 (Joseph A. Walker and

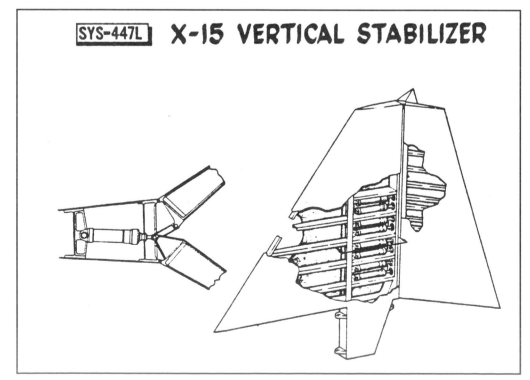

The vertical stabilizer was one of the most obvious changes between the industry conference configuration and the final vehicle. The first design did not use the exaggerated wedge-shape of the final unit. It was also more traditional, using a fixed forward portion and a conventional appearing rudder. The final version used an all-moving design. Note the rudder splits to become speed brakes, much like the shuttle design 25 years later. (NASA)

Iven C. Kincheloe, Jr.[24]), it would appear that the planned controllers were not all that might have been desired.

A request that the pilot be provided with continuous information on the nose-wheel door position (loss of the door could produce severe structural damage) was rejected because the committee felt that the previously approved suggestion for gear-up inspection panels would make such information unnecessary. This particular item would come back to haunt the program during the flight research phase.

After the completion of the development engineering inspection, the X-15 airframe design changed only in relatively minor details. North American essentially built the X-15 described at the industry conference in October and inspected in mockup in December 1956. Continued wind tunnel testing resulted in some external modifications, particularly of the vertical stabilizer, and some weight changes occurred as plans became more definite. But while work on the airframe progressed smoothly, with few unexpected problems, the project as a whole did encounter difficulties, some of them serious enough to threaten long delays. In fact, North American's rapid preparation of drawings and production planning served to highlight the lack of progress on some of the components and subsystems that were essential to the success of the program.

The Engine

Those concerned with the success of the X-15 had to monitor the development of the aircraft itself, the XLR99 rocket engine, the auxiliary power units, an inertial system, a tracking range, a pressure suit, and an ejection seat. They had to make arrangements for support and B-36 carrier aircraft, ground equipment, the selection of pilots, and the development of simulators for pilot training. It was necessary to secure time on centrifuges, in wind tunnels, and on sled tracks. The ball-nose had to be developed, studies made of the compatibility

of the X-15 and the carrier aircraft, and other studies on the possibility of extending the X-15 program beyond the goals originally contemplated. In addition to such tasks, funds to cover ever increasing costs had to be secured if the project were to have any chance of ultimate success, and at certain stages the effects of possibly harmful publicity had to be considered. With such multiplicity of tasks, it could be expected that several serious problems would arise; not surprisingly, probably the most serious arose during the development of the XLR99.

Finding a suitable engine for the X-15 had been somewhat problematic from the earliest stages of the project, when the WADC Power Plant Laboratory had pointed out that the lack of an acceptable rocket engine was the major shortcoming of the NACA's original proposal. The laboratory did not believe that any available engine was entirely suitable for the X-15 and held that no matter what engine was accepted, a considerable amount of development work could be anticipated. Most of the possible engines were either too small or would need too long a development period. In spite of these reservations, the laboratory listed a number of engines worth considering and drew up a statement of the requirements for an engine that would be suitable for the proposed X-15 design. The laboratory also made clear its stand that the government should "… accept responsibility for development of the selected engine and … provide this engine to the airplane contractor as Government Furnished Equipment."[25]

The primary requirement for an X-15 engine, as outlined in 1954, was that it be capable of operating safely under all conditions. Service life would not have to be as long as for a production engine, but engineers hoped that the selected engine would not depart too far from production standards. The same attitude was taken toward reliability; the engine need not be as reliable as a production article, but it should approach such reliability as nearly as possible. There could be no altitude limitations for starting or operating the engine, and the power plant would have to be entirely safe during start, operation, and shutdown, no matter what the altitude. The laboratory made it quite clear that a variable thrust engine capable of repeated restarts was essential.

The engine ultimately selected was not one of the four originally presented as possibilities by the Power Plant Laboratory. The ultimate selection was foreshadowed, however, in discussions with Reaction Motors concerning the XLR10, during which attention was drawn to what was termed "... a larger version of [the] Viking engine [XLR30]." In light of subsequent events, it was interesting to note that the laboratory thought[26] the XLR30 could be developed into a suitable X-15 engine for "... less than $5,000,000 ..." and with " ... approximately two years' work."[27]

After North American had been selected as the winner of the X-15 competition, plans were instituted to procure the modified XLR30 engine that had been incorporated in the winning design. Late in October, Reaction Motors was notified that North American had won the X-15 competition and

that the winner had based his proposals upon the XLR30 engine.[28]

On 1 December 1955 a $1,000,000 letter contract was initiated with Reaction Motors for the development of a rocket engine for the X-15.[29] Soon afterwards, a controversy developed over the assignment of cognizance for the development of the engine. It began with a letter from Rear Adm. W. A. Schoech of the Bureau of Aeronautics. Adm. Schoech contended that since the XLR30-RM-2 rocket engine was the basis for the X-15 power plant, and the BuAer had already devoted three years to the development of that engine, it would be logical to assign the responsibility for further development to the Navy. The admiral felt that retention of the program by the BuAer would expedite development, especially as the Navy could direct the development toward an X-15 engine by making specification changes rather than by negotiating a new contract.[30]

The Navy's bid for control of the engine development was rejected on 3 January 1956 on the grounds that the management responsibility should be vested in a single agency, that conflict of interest might generate delay,

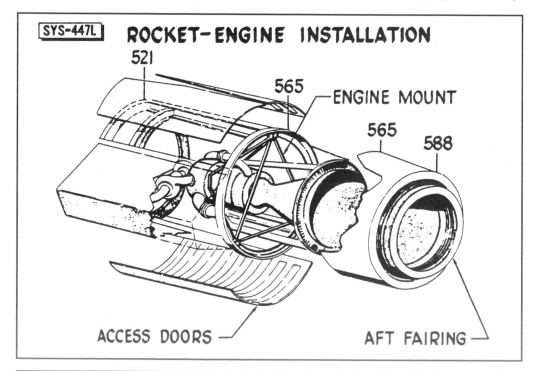

SYS-447L ROCKET-ENGINE INSTALLATION

521

565 — ENGINE MOUNT

565

588

ACCESS DOORS

AFT FAIRING

The XLR99 was an extremely compact engine, considering it was able to produce over 57,000 pounds-thrust. This was the first throttleable and restartable man-rated rocket engine. Many of the lessons-learned from this engine were incorporated into the Space Shuttle Main Engine developed 20 years later. (NASA)

and that BuAer was underestimating the time and effort necessary to make the XLR30 a satisfactory engine for piloted flight.

The final Reaction Motors technical proposal was received by the Power Plant Laboratory on 24 January, with the cost proposal following on 8 February.[31] The cover letter from Reaction Motors promised delivery of the first complete system "… within thirty (30) months after we are authorized to proceed."[32] Reaction Motors also estimated that the entire cost of the program would total $10,480,718.[33] On 21 February the new engine was designated XLR99-RM-1.[34]

The 1956 industry conference heard two papers on the proposed engine and propulsion system for the X-15. The XLR99-RM-1 would be able to vary its thrust from 19,200 to 57,200 pounds at 40,000 feet using anhydrous ammonia and liquid oxygen (LOX)[35] as propellants. Specific impulse was to vary from a minimum of 256 seconds to a maximum of 276 seconds. The engine was to fit into a space 71.7 inches long and 43.2 inches in diameter, have a dry weight of 618 pounds, and a wet weight of 748 pounds. A single thrust chamber was supplied by a

hydrogen-peroxide-driven turbopump, with the turbopump's exhaust being recovered in the thrust chamber. Thrust control was by regulation of the turbopump speed.[36]

The use of ammonia as a propellant presented some potential problems; in addition to being toxic in high concentrations, ammonia is also corrosive to all copper-based metals. There were discussions early in the program between the Air Force, Reaction Motors, and the Lewis Research Center[37] about the possibility of switching to a hydrocarbon fuel. It was finally concluded that changing fuel would add six months to the development schedule; it would be easier to learn to live with the ammonia.[38] There is no documentation that the ammonia ultimately presented any significant problems to the program.

The decision to control thrust by regulating the speed of the turbopump was made because the other possibilities (regulation by measurement of the pressure in the thrust chamber or of the pressure of the discharge) would cause the turbopump to speed up as pressure dropped. As the most likely cause of pressure drop would be cavitation in the propellant system, an increase in turbopump

This 1956 sketch shows the controls and indicators for the XLR99. A different set of controls were used for the XLR11 flights, although they fit into the same space allocation. Notice the simple throttle on the left console, underneath the reaction control side-stick (not shown). The jettison controls took on particular significance on missions that had to be aborted prior to engine burn-out. (NASA)

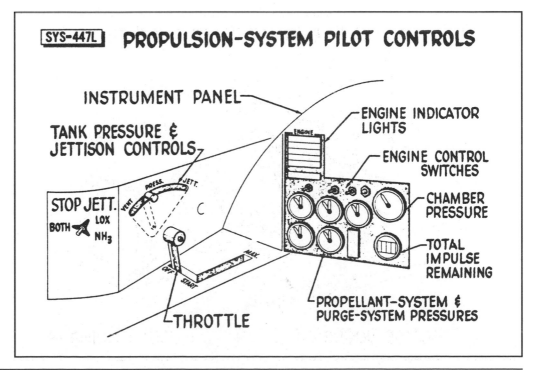

speed would aggravate rather than correct the situation. Reaction Motors had also decided that varying the injection area was too complicated a method for attaining a variable thrust engine and had chosen to vary the injection pressure instead.

The regenerative cooling of the thrust chamber created another problem since the variable fuel flow of a throttleable engine meant that the system's cooling capacity would also vary and that adequate cooling throughout the engine's operating range would produce excessive cooling under some conditions. Engine compartment temperatures also had to be given more consideration than in previous rocket engine designs because of the higher radiant heat transfer from the structure of the X-15. Reaction Motors' spokesman at the 1956 industry conference concluded that the development of the XLR99 was going to be a difficult task. Subsequent events were certainly to prove the validity of that prediction.

A second paper dealt with engine and accessory installation, the location of the propellant system components, and the engine controls and instruments. The main propellant tanks were to contain the LOX, ammonia, and the hydrogen peroxide. The LOX tank,

with a capacity of approximately 1,000 gallons, was located just ahead of the aircraft's center of gravity; the 1,400 gallon ammonia tank was just aft of the same point. A center core tube within the LOX tank would provide a location for a supply of helium under a pressure of 3,600 psi. Helium was used to pressurize both the LOX and ammonia tanks. A 75-gallon hydrogen peroxide tank behind the ammonia tank provided the monopropellant for the turbopump.

Provision was also made to top-off the LOX tank from a supply carried aboard the carrier aircraft; this was considered to be beneficial in two ways. The LOX supply in the carrier aircraft could be kept cooler than the oxygen already aboard the X-15, and the added LOX would permit cooling of the X-15's own supply by boil-off, without reduction of the quantity available for flight. The ammonia tank was not to be provided with a top-off arrangement, as the slight increase in fuel temperature during flight was not considered significant enough to justify the complications such a system would have entailed.

On 10 July 1957, Reaction Motors advised the Air Force that an engine satisfying the contract specifications could not be developed unless the government agreed to a nine-

The XLR99 on a maintenance stand. The engine used ammonia (NH3) as fuel and liquid oxygen (LOX) as the oxidizer. The XLR99 required a separate propellant, hydrogen peroxide, to drive its high-speed turbopump—the Space Shuttle Main Engine uses the propellant itself (LH2 or LO2, as appropriate) to drive the turbopumps. (AFFTC via the Tony Landis Collection)

month schedule extension and a cost increase from $15,000,000 to $21,800,000. At the same time, Reaction Motors indicated that it could provide an engine that met the performance specification within the established schedule if permitted to increase the weight from 618 pounds to 836 pounds. The company estimated that this overweight engine could be provided for $17,100,000. The Air Force elected to pursue the heavier engine since it would be available sooner and have less impact on the overall X-15 program.

Those who hoped that the overall performance of the X-15 would be maintained were encouraged by a report that the turbopump was more efficient than anticipated and would allow a 197 pound reduction in the amount of hydrogen peroxide necessary for its operation. This decrease, a lighter than expected airframe, and the increase in launch speeds and altitudes provided by a recent substitution of a B-52 as the carrier aircraft, offered some hope that the original X-15 performance goals might still be achieved.[39]

Despite the relaxation of the weight requirements, the engine program failed to proceed at a satisfactory pace. On 11 December 1957 Reaction Motors reported a new six-month slip. The threat to the entire X-15 program posed by these new delays was a matter of serious concern, and on 7 January 1958, Reaction Motors was asked to furnish a detailed schedule and to propose means for solving the difficulties. The new schedule, which reached WADC in mid-January, indicated that the program would be delayed another five and one-half months and that costs would rise to $34,400,000—double the cost estimate of the previous July.[40]

In reaction, the Air Force recommended increasing the resources available to Reaction Motors and proposed the use of two [41] XLR11 rocket engines as an interim installation for the initial X-15 flights. Additional funds to cover the increased effort were also approved, as was the establishment of an advisory group.[42]

The threat that engine delays would seriously impair the value of the X-15 program had generated a whole series of actions during the first half of 1958: personal visits by general officers to Reaction Motors, numerous conferences between the contractor and representatives of government agencies, increased support from the Propulsion Laboratory[43] and the NACA, an increase in funds, and letters containing severe censure of the company's conduct of the program. An emergency situation had been encountered, emergency remedies were used, and by midsummer improvements began to be noted.

Engine progress continued to be reasonably satisfactory during the remainder of 1958. A destructive failure that occurred on 24 October was traced to components that had already been recognized as inadequate and were in the process of being redesigned. The failure, therefore, was not considered of major importance.[44] A long-sought goal was finally reached on 18 April 1959 with completion of the Preliminary Flight Rating Test (PFRT). The flight rating program began at once.[45]

At the end of April, a "realistic" schedule for the remainder of the program showed that the Flight Rating Test would be completed by 1 September 1959. The first ground test engine was delivered to Edwards AFB at the end of May, and the first flight engine was delivered at the end of July.[46]

A total of 10 flight engines were initially procured, along with six spare injector-chamber assemblies; one additional flight engine was subsequently procured. In January 1961, shortly after the first XLR99 test flight, only eight of these engines were available to the flight test program. There was still a number of problems with the engines that Reaction Motors was continuing to work on; the most serious being a vibration at certain power levels, and a shorter than expected chamber life. There were four engines being used for continued ground tests, including two flight engines.[47] Three of the engines were involved in tests to isolate and eliminate the vibrations, while the fourth engine was being used to investigate extending the life of the chamber.[48]

It is interesting to note that early in the proposal stage, North American determined that aerodynamic drag was not as important a design factor as was normally the case with jet-powered fighters. This was largely due to the amount of excess thrust available from the XLR99. Weight was considered a larger driver in the overall airplane design. Only about 10 percent of the total engine thrust was necessary to overcome drag, and another 20 percent to overcome weight. The remaining 70 percent of engine thrust was available to accelerate the X-15.[49]

Other Systems

In early 1958, at the very height of the furor over the problems with the XLR99, a note of warning sounded for the General Electric auxiliary power unit (APU). On 26 March 1958 and again on 11 April 1958, General Electric notified North American of its inability to meet the original specifications in the time available, and requested approval

of new specifications. North American, with the concurrence of the Air Force, agreed to modify the requirements. The major changes involved an increase in weight from 40 to 48 pounds, an increase in start time from five to seven seconds, and a revision of the specific fuel consumption curves.[50]

By the end of the summer 1958, the APU seemed to have reached a more satisfactory state of development, and production units were ready for shipment.[51] The early captive flights beginning in 1959 would reveal some additional problems, but investigation showed that the in-flight failures had occurred partially because captive testing subjected the units to an abnormal operational sequence that would not be encountered during glide and powered flight. Some components were redesigned, but the APU would continue to be relatively troublesome in actual service.

During the course of the X-15 program, many concerns were voiced over the development of a pressure suit and an escape system. Although full-pressure suits had been studied during World War II, attempts to fabricate a practical garment had met with failure. The

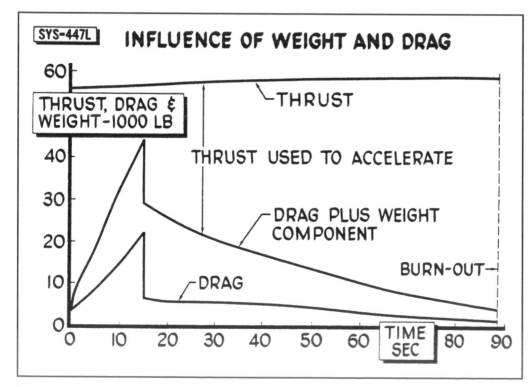

Soulé to Storms: "You have a little airplane and a big engine with a large thrust margin."

And indeed they did. The XLR99 provided 57,000 pounds-thrust to propel an aircraft that only weighed 30,000 pounds. Consider that the contemporary F-104 Starfighter, considered something of a hot rod, weighed 20,000 pounds and its J79 only produced 15,000 pounds-thrust in full afterburner. (NASA)

Air Force took renewed interest in pressure suits in 1954 when it had become obvious that the increasing performance of aircraft was going to necessitate such a garment. The first result of the renewed interest was the creation of a suit that was heavy, bulky, and unwieldy; the garment had only limited mobility and various joints created painful pressure points. However, in 1955 the David Clark Company succeeded in producing a garment using a distorted-angle fabric that held some promise of ultimate success.[52]

Despite the early state-of-development of full-pressure suits, Scott Crossfield was convinced they were the way to go for X-15. North American's detail specifications of 2 March 1956 called for just such a garment—to be furnished by North American through a subcontract with the David Clark Company.[53] A positive step toward Air Force acceptance of the idea occurred during a conference held at the North American plant on 20-22 June 1956. A full-pressure suit developed by the Navy was demonstrated during an inspection of the preliminary cockpit mockup, and although the suit still had a number of deficiencies, it was concluded that "... the state-of-the-art on full pressure suits should permit the development of such a suit satisfactory for use in the X-15."[54]

After an extremely difficult and prolonged development process, Scott Crossfield received the first new MC-2 full-pressure suit on 17 December 1958 and, two days later, the suit successfully passed nitrogen contamination tests at the Air Force Aero Medical Laboratory. The X-15 project officer attributed much of the credit for the successful and timely qualification of the full-pressure suit to the intensive efforts of Crossfield.[55]

Fortunately, development did not stop there. On 27 July 1959, the Aero Medical Laboratory brought the first of the new A/P22S-2 pressure suits to Edwards. The consensus amongst the pilots was that it represented a large improvement over the earlier MC-2. It was more comfortable and pro-

vided greater mobility; and it took only 5 minutes to put on, compared to 30 minutes for the MC-2. However, it would take another year before fully-qualified versions of the suit were delivered to the X-15 program.[56]

While not directly related to the pressure suit difficulties, the type of escape system to be used in the X-15 had been the subject of debate at an early stage of the program; the decision to use the stable-seat, full-pressure-suit combination had been a compromise based largely on the fact that the ejection seat was lighter and offered fewer complications than the other alternatives.

As early as 8 February 1955, the Aero Medical Laboratory had recommended a capsular escape system, but the laboratory had also admitted that such a system would probably require extensive development. The second choice was a stable seat that incorporated limb retention features and one that would produce a minimum of deceleration.[57] During meetings held in October and November 1955, it was agreed that North American would design an ejection seat for the X-15 and would also prepare a report justifying the use of such a system in preference to a capsule. North American was to incorporate head and limb restraints in the proposed seat.[58]

Despite the report, the Air Force was not completely convinced. At a meeting held at Wright Field on 2-3 May 1956, the Air Force again pointed out the limitations of ejection seats. In the opinion of one NACA engineer who attended the meeting, the Air Force was still strongly in favor of a capsule—partly because of the additional safety a capsule system would offer, and partly because the use of such a system in the X-15 would provide an opportunity for further developmental research. Primarily due to the efforts of Scott Crossfield, the participants finally agreed that because of the "time factor, weight, ignorance about proper capsule design, and the safety features being built into the airplane structure itself, the X-15 was probably its own best capsule." About the only result of the reluctance of the Air Force to endorse an ejection seat was a request that North American yet again document the arguments for the seat.[59]

The death of Captain Milburn G. Apt in the crash of the Bell X-2, which had been equipped with an escape capsule, in September 1956 renewed apprehension as to the adequacy of the X-15's escape system.[60] By this time, however, it was acknowledged that no substantive changes could be made to the X-15 design. Fortunately, North American's seat development efforts were generally proceeding well.[61]

Sled tests of the ejection seat began early in 1958 at Edwards with the preliminary tests concluded on 22 April. Because of the high cost of sled runs, the X-15 project office advised North American to eliminate the planned incremental testing and to conduct the tests at just two pressure levels—125 pounds per square foot and 1,500 pounds per square foot. The X-15 office felt that successful tests at these two levels would furnish adequate proof of seat reliability at intermediate pressures.[62]

Between 4 June 1958 and 3 March 1959, the X-15 seat completed its series of sled tests. Various problems, with both the seat and the sled, had been encountered, but all had been worked through to the satisfaction of North American and the Air Force. The X-15 seat was cleared for flight use.[63]

Another item for which the Air Force retained direct responsibility was the all-attitude inertial flight data system. It was realized from the beginning of the X-15 program that the airplane's performance would necessitate a new means of determining altitude, speed, and aircraft attitude. This was because the traditional use of static pressure as a reference would be largely impossible at the speeds and altitudes the X-15 would achieve; moreover, the temperatures encountered would rule out the use of tradition pitot tube sensing devices. The NACA had proposed a "stable-platform iner-tial-integrating and attitude sensing unit" as the means of meeting these needs.[64] A series of miscommunications resulted in the NACA assuming the Air Force had already developed a satisfactory unit and would provide it to the X-15 program.[65] After it was discovered that a suitable unit did not exist, emergency efforts were undertaken to develop one without impacting the X-15 program. After a considerable amount of controversy, a sole-source contract was awarded to the Sperry Gyroscope Company on 5 June 1957 for the development and manufacture of the stable-platform.[66] The cost-plus-fixed-fee contract, signed on 5 June 1957, was for $1,213,518.06 with a fixed fee of $85,000.[67]

In April 1958, the Air Force concluded that the scheduled delivery of the initial Sperry unit in December would not permit adequate testing to be performed prior to the first flights of the X-15. Consequently a less capable interim gyroscopic system was installed in the first two aircraft and the final Sperry system was installed in the last X-15.[68]

By the end of 1958, the two major system components (the stabilizer and the computer) were completed and ready to be tested as a complete unit. The systems were shipped to Edwards in late January 1959, and during the spring of 1959 plans were made to use the NB-52 carrier aircraft as a test vehicle.[69] In addition, arrangements were made to test the stable-platform in a KC-97 that was already in use as a test aircraft in connection with the B-58 program.[70] The first test flights in the KC-97 were carried out in late April.[71] By June, North American had successfully installed the Sperry system in the third X-15.[72] In January 1961, wiring was installed in the NB-52B to allow the stable-platform to be installed in a pod carried on the pylon under the wing. The first complete stable-platform system installed in the B-52 pod was flown on 1 March 1961. Since the B-52 was capable of greater speeds and higher altitudes than the KC-97, it provided additional data to assist Sperry in resolving problems being encountered with the unit.[73]

The High Range

Previous rocket aircraft, such as the X-1 and X-2, had been able to conduct the majority of their flight research in the skies directly over the Edwards test areas. The capabilities of the X-15, however, would use vastly more airspace. The proposed trajectories required an essentially straight flight corridor equipped with multiple tracking, telemetry, and communications sites, as well as the need for suitable emergency landing areas. This led to construction of the X-15 High Range extending from Wendover, Utah, to Edwards AFB. Radar and telemetry stations were installed at Ely and Beatty, Nevada, as well as Edwards. Telemetry from the X-15, as well as voice communications, were received, recorded, and forwarded to Edwards by the stations at Ely and Beatty. Each of these stations was also manned by a person to back up the prime "communicator" (NASA 1) at Edwards in case the communication links went down. Each ground station overlapped the next, and they were interconnected via microwave and land-line so that timing signals, voice communication, and radar data would be available to all. Provisions were made for recording the acquired data on tape and film, although some of the data was directly displayed on strip and plotting charts. The design and construction of the range was accomplished by Electronic Engineering Company of Los Angeles under contract with the Air Force.[74] North American and the NACA also conducted numerous evaluations of various dry lakes to determine which were suitable for emergency landings along the route (see the summary included as an appendix to this monograph).

Carrier Aircraft

The group at Langley had sized their X-15 proposal around the potential of using a

The use of a B-36 carrier aircraft would have allowed the pilot to exit the aircraft while in transit to the drop area, or in case of emergency. However, personnel at the FRC worried that the B-36 would not be supportable since it was being phased out of active service. In the end, the B-52 provided much better performance and was ultimately selected. (AFFTC History Office)

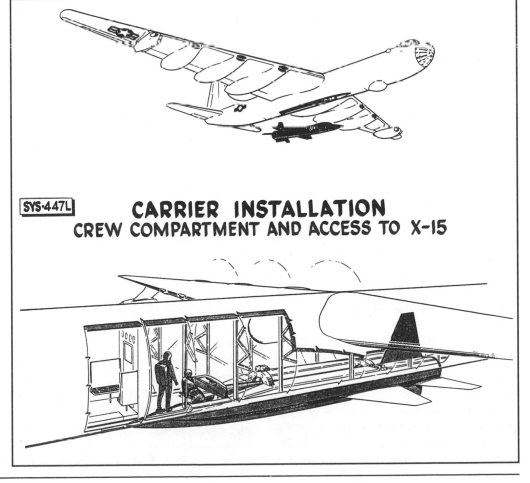

SYS-447L
CARRIER INSTALLATION
CREW COMPARTMENT AND ACCESS TO X-15

Convair B-36 as the carrier aircraft. This was a natural extension of previous X-planes that had used a Boeing B-29 or B-50 as a carrier. The B-36 would be modified to carry the X-15 partially enclosed in its bomb bays, much like the X-1 and X-2 had been in earlier projects. This arrangement had some advantages; the pilot could freely move between the X-15 and B-36 during climb-out and the cruise to the launch location. This was extremely advantageous if problems developed that required jettisoning the X-15 prior to launch. At the time of the first industry conference in 1956, it was expected that a B-36 would be modified beginning in the middle of 1957 and be ready for flight tests in October 1958.[75]

As the weight of the X-15 and its subsystems grew, however, the Air Force and NASA began to look for ways to recover some of the lost performance. One way was to launch the X-15 at a higher altitude and greater speed. In addition, the personnel at Edwards believed that the ten-engine B-36 would be difficult to maintain[76] since it was being phased out of the Air Force inventory. Investigations showed that the X-15, as designed, would fit under the wing of one of the new Boeing B-52 Stratofortresses; the configuration of the B-52 precluded carrying the X-15 in the bomb bay. This was not the ideal solution—the X-15 pilot would have to be locked in the research airplane prior to takeoff, and the large weight transition when the X-15 was released would provide some interesting control problems for the B-52. Further analysis concluded that the potential problems were solvable, and that the increase in speed and altitude capabilities were desirable. Fortunately, two early B-52s were completing their test duties, and the Air Force made them available to the program.

On 29 November 1957, the B-52A (52-003) arrived at Air Force Plant 42 in Palmdale, California, after a flight from the Boeing plant in Seattle. The aircraft was placed in storage pending modifications. On 4 February 1958, the B-52A was moved into the North American hanger at Plant 42 and modified with a large pylon under the wing, the capa-

bility to monitor to the X-15, and a system to replenish the X-15 LOX supply. The aircraft, now designated[77] NB-52A, was flown to Edwards AFB on 14 November 1958; it was later named "The High and the Mighty One." The Air Force also supplied a B-52B (52-008) that arrived in Palmdale for similar modifications on 5 January 1959, and was flown, as an NB-52B, to Edwards on 8 June 1959.

Roll Out

As the first X-15 was being completed, the NACA held the second X-15 industry conference in Los Angeles on 28-29 July 1958. North American began the conference with a paper detailing the developmental status of the aircraft. Twenty-seven other papers covered subjects such as stability and control, simulator testing, pilot considerations, mission instrumentation, thermodynamics, structures, materials and fabrication. There were approximately 550 attendees.[78]

On 1 October 1958, High-Speed Flight Station employees Doll Matay and John Hedgepeth put up a ladder in front of the station building at the foot of Lilly Avenue and took down the winged-shield NACA emblem from over the entrance door. NASA had arrived in the desert, bringing with it a new era of space-consciousness, soaring budgets, and publicity. The old NACA days of concentration on aeronautics, and especially aerodynamics, were gone forever, as was the agency itself. On this day, the National Aeronautics and Space Administration was created.[79]

The X-15 construction process eventually consumed just over two years, and on 15 October 1958, the first aircraft (56-6670) was rolled out. Following conclusion of the official ceremonies, it was moved back inside and prepared for shipment to Edwards. On the night of 16 October, covered completely in protective heavy-duty wrapping paper, it was shipped by truck to Edwards for initial ground test work.

Chapter 2
Notes and
References

1   Letter from Scott Crossfield to Dennis R. Jenkins, 30 June 1999.

2   Harrison A. Storms, "X-15 Hardware Design Challenges" (a paper in the *Proceedings of the X-15 30th Anniversary Celebration*, Dryden Flight Research Facility, Edwards, California, 8 June 1989, NASA CP-3105), p. 27.

3   Letter from Scott Crossfield to Dennis R. Jenkins, 30 June 1999.

4   The High-Speed Flight Station (HSFS) became the Flight Research Center (FRC) on 27 September 1959, and the Hugh L. Dryden Flight Research Center (usually abbreviated DFRC) on 26 March 1976.

5   Wendell H. Stillwell, *X-15 Research Results*, (NASA, Washington, DC.: NASA SP-60, 1965).

6   James E. Love, "History and Development of the X-15 Research Aircraft," not dated, DFRC History Office, p. 10.

7   Eventually, an Air Force-NACA study team journeyed to France to study the prototype Sud-Ouest Trident interceptor, which had such a tail configuration. See "Beyond the Frontiers, Sub-Quest Trident: Mixed-Powerplant Fighter," Wings of Fame, Aerospace Publishing Ltd., London, Volume 10, p. 32.

8   Memorandum from M. A. Todd, Acting Chief, Contractor Reporting and Bailment Branch, Support Division, to Chief, Fighter Branch, Aircraft Division, Director Procurement and Production, AMC, 15 June 1956, subject: Confirmation of Serial Numbers Assigned, in the files of the AFMC History Office, Wright-Patterson AFB (WPAFB), Ohio.

9   Letter from Dr. Hugh L. Dryden, Director of NACA, to Chief, Fighter WSPO, Director of Systems Management, ARDC, 6 July 1956, no subject, in the files of the AFMC History Office, WPAFB, Ohio.

10   *Research Airplane Committee Report on the Conference on the Progress of the X-15 Project*, a compilation of the papers presented at the Langley Aeronautical Laboratory, Langley Field, Virginia, 25-26 October 1956.

11   Ibid., pp. 23-31.

12   The list included facilities at Langley, Ames, North American, and the Massachusetts Institute of Technology.

13   Dale L. Compton, "Welcome," (a paper in the *Proceedings of the X-15 30th Anniversary Celebration*, Dryden Flight Research Facility, Edwards, California, 8 June 1989, NASA CP-3105), p. 3.

14   The interim XLR11 engine would use liquid oxygen as the oxidizer and an ethyl alcohol-water mixture as fuel.

15   S. A. Sjoberg, "Some Experience With Side Controllers" (a paper presented at the NACA Conference on the Progress of the X-15 Project, Langley Aeronautical Laboratory, Langley Field, Virginia, 25-26 October 1956), pp. 167-171.

16   X-15 Interim Flight Manual, FHB-23-1, 18 March 1960, changed 12 May 1961.

17   Today this would be called a reaction control system.

18   Also known as the Q-ball.

19   A gimbal allows a body to incline in predefined directions. In this case the sphere could move both left-right and up-down in relation to the nose. Electric servomotors (servos) provided the power to move the sphere as necessary.

20   Memorandum from Arthur W. Vogeley, Aeronautical Research Scientist, NACA, to Research Airplane Project Leader, Langley Aeronautical Laboratory, NACA, 30 November 1955, subject: Project 1226 meetings to discuss changes in the North American Proposal—Wright-Patterson AFB meeting of 24-25 October, and North American Aviation meetings in Los Angeles on 27-28 October and 14-15 November 1955.

21   Walter C. Williams, "X-15 Concept Evolution" (a paper in the *Proceedings of the X-15 30th Anniversary Celebration*, Dryden Flight Research Facility, Edwards, California, 8 June 1989, NASA CP-3105), p. 13.

22   Lawrence P. Greene, "Summary of Pertinent Problems and Current Status of the X-15 Airplane" (a paper presented at the NACA Conference on the Progress of the X-15 Project, Langley Aeronautical Laboratory, Langley Field, Virginia, 25-26 October 1956), pp. 239-258.

23   A "request for alteration" is the form used to request changes as the result of a mockup inspection within the Air Force.

24   Kinchloe had been selected as the Air Force project pilot for the X-15 program in September 1957. Unfortunately, he was killed while attempting to eject from an F-104 at Edwards on 26 July 1958.

25   Memorandum from T. J. Keating, Chief, Non-Rotating Engine Branch, Power Plant Laboratory, to Chief, New Development Office, 15 November 1954, in the files of the AFMC History Office, WPAFB, Ohio.

26   In fairness to the laboratory, it must be admitted that such estimates were accompanied by a statement that "… less confidence in these estimates exists because the XLR30 engine is at present in a much earlier stage of development." This qualification proved to have been justified.

27   Memorandum from T. J. Keating, Chief, Non-Rotating Engine Branch, Power Plant Laboratory, to Chief, New Development Office, 15 November 1954, in the files of the AFMC History Office, WPAFB, Ohio.

28   Letter from Lieutenant H. J. Savage, Power Plant Laboratory, WADC, to Reaction Motors, Inc., 26 October 1955, subject: Engine for the X-15 Airplane, in the files of the AFMC History Office, WPAFB, Ohio.

29   Letter from Lieutenant C. E. McCollough Jr., New Development Office, Director of WSO, WADC to Chief, Non-Rotating Engine Branch, 1 December 1955, in the files of the AFMC History Office, WPAFB, Ohio.

30   Letter from Rear Adm. W. A. Schoech, Assistant Chief for Research and Development, BuAer, USN, to C/S, USAF, 28 November 1955, subject: Cognizance over development of rocket power plant for NACA X-15 research airplane, in the files of the AFMC History Office, WPAFB, Ohio.

31   Letter from Brigadier General V. R. Haugen, Deputy Commander for Development, WADC, to NACA-Washington, 15 February 1956, subject: Engine Contract for the X-15, in the files of the AFMC History Office, WPAFB, Ohio.

32   Letter from W. P. Turner, Manager, Contracts Division, Reaction Motors, Inc., to Commander, AMC, 7 February 1956, subject: Rocket Engine System for X-15 Research Aircraft, in the files of the AFMC History Office, WPAFB, Ohio

33   Letter from W. P. Turner, Manager, Contracts Division, Reaction Motors, Inc., to Commander, AMC, 7 February 1956, subject: Rocket Engine System for X-15 Research Aircraft, in the files of the AFMC History Office, WPAFB, Ohio.

34   The designation became "official" at Wright-Field on 6 March and received Navy approval on 29 March. Many documents, particularly later in the flight program, list the designation as YLR99, but no evidence was discovered during research for this monograph that indicated this was ever an official designation.

35   Today liquid oxygen is usually abbreviated LO2, but it was common practice up until the mod-1980s to use LOX.

36   The engine was eventually to undergo numerous changes of detail but its basic design, as described to the conference, was not greatly altered.

[37]   The Aircraft Engine Research Laboratory was founded on 23 June 1941; in April 1947 it was renamed the Flight Propulsion Research Laboratory. A year later it was renamed the Lewis Flight Propulsion Laboratory. When NASA was formed on 1 October 1958, the laboratory was renamed the Lewis Research Center (abbreviated LeRC to differentiate it from the Langley Research Center—LaRC). On 1 March 1999 it was renamed the John H. Glenn Research Center at Lewis Field.

[38]   Walter C. Williams, "X-15 Concept Evolution" (a paper in the *Proceedings of the X-15 30th Anniversary Celebration*, Dryden Flight Research Facility, Edwards, California, 8 June 1989, NASA CP-3105), p. 15.

[39]   Memorandum from Arthur W. Vogeley to Research Airplane Project Leader, NACA, 3 August 1957, subject: X-15 Airplane—Discussions at Air Research and Development Command, Detachment #1, Wright-Patterson Air Force Base, Dayton, Ohio, on 29-30 July 1957, in the files of the AFMC History Office, WPAFB, Ohio.

[40]   Status Report of the XLR99-RM-1, 9 January 1958 through 27 June 1958, Prepared by Propulsion Laboratory, WADC, in the files of the AFMC History Office, WPAFB, Ohio.

[41]   There was a clear distinction between proposals for an interim engine to permit flight trials before an XLR99 became available, and an alternate engine, to substitute for the XLR99 in the final X-15.

[42]   Interview of C. E. McCollough, 14 May 1959, by R. S. Houston, in the files of the Air Force Museum archives.

[43]   The Power Plant and Propeller laboratories had been combined on 17 June 1957 into the Propulsion Laboratory.

[44]   X-15 WSPO Weekly Activity Report, 14 November 1958, in the files of the AFMC History Office, WPAFB, Ohio.

[45]   X-15 WSPO Weekly Activity Report, 24 April 1959, in the files of the AFMC History Office, WPAFB, Ohio.

[46]   X-15 WSPO Weekly Activity Report, 8 May 1959, in the files of the AFMC History Office, WPAFB, Ohio.

[47]   The flight engines were s/n 101 and 102; s/n 6 and 12 were the dedicated ground engines.

[48]   James E. Love, "History and Development of the X-15 Research Aircraft," not dated, DFRC History Office, p. 18.

[49]   Charles H. Feltz, "Description of the X-15 Airplane, Performance, and Design Mission" (a paper presented at the NACA Conference on the Progress of the X-15 Project, Langley Aeronautical Laboratory, 25-26 October 1956), pp. 28.

[50]   X-15 WSPO Weekly Activity Report, 2 May 1958, in the files of the AFMC History Office, WPAFB, Ohio.

[51]   X-15 WSPO Weekly Activity Report, 5 September 1958, in the files of the AFMC History Office, WPAFB, Ohio.

[52]   Research Airplane Committee, Report on Conference on the Progress of the X-15 Project, compilation of the papers presented at the IAS Building, Los Angeles, California, 28-30 July 1958, in files of X-15 WSPO, pp. 117.

[53]   Report, *Detail Specifications NA5-4047*, 2 March 1956, in the files of the AFMC History Office, WPAFB, Ohio.

[54]   X-15 WSPO Weekly Activity Report, 28 June 1956, in the files of the AFMC History Office, WPAFB, Ohio.

[55]   X-15 WSPO Weekly Activity Report, 30 January 1959; in the files of the AFMC History Office, WPAFB, Ohio.

[56]   James E. Love, "History and Development of the X-15 Research Aircraft," not dated, DFRC History Office, p. 13.

[57]   Memorandum from H. E. Savely, Chief, Biophysics Branch, Aero Medical Laboratory, WADC, to Chief, New Development Office, Fighter Aircraft Division, Director of WSO, 8 February 1955, subject: Acceleration Tolerance and Emergency Escape, in the files of the AFMC History Office, WPAFB, Ohio.

[58]   Memorandum from Arthur W. Vogeley, Aeronautical Research Scientist, NACA, to Research Airplane Project Leader, 30 November 1955, in the files of the AFMC History Office, WPAFB, Ohio.

[59]   Memorandum from Hartley Soulé to Members, NACA Research Airplane Project Panel, 7 June 1956, in the files of the NASA History Office.

[60]   Memorandum from Brigadier General M. C. Demler, Deputy Commander for R&D, to Deputy Commander for Weapons Systems, ARDC, 2 January 1957, subject: Escape Systems for Research Vehicles such as the X-15, in the files of the AFMC History Office, WPAFB, Ohio.

[61]   Letter from R. H. Rice, Vice President and General Manager, North American Aviation, to Commander, AMC, 31 January 1957, subject: Contract AF33(600)-31693, X-15 Airplane, GFAE Ejection Seat Catapult—Change to CFE Ballistic Rocket Type-ECP NA-X-15-19, in the files of the AFMC History Office, WPAFB, Ohio.

[62]   X-15 WSPO Weekly Activity Report, 21 May 1958, in the files of the AFMC History Office, WPAFB, Ohio.

[63]   X-15 WSPO Weekly Activity Report, 13 March 1959, in the files of the AFMC History Office, WPAFB, Ohio.

[64]   Today, this is more often referred to as an inertial measurement unit, similar to what forms the basis of most inertial navigation systems.

[65]   Memorandum from Walter C. Williams to Research Airplane Project Leader, 27 January 1956, in the files of the AFMC History Office, WPAFB, Ohio.

[66]   Proposal Number A. E. 1752, "Development of Flight Research Stabilized Platform," Sperry Gyroscope Co.(contract AF33-600-35397), in the files of the AFMC History Office, WPAFB, Ohio.

[67]   Contract AF33(600)-35397, 5 June 1957, in the files of the AFMC History Office, WPAFB, Ohio.

[68]   X-15 WSPO Weekly Activity Report, 2 May 1958, in the files of the AFMC History Office, WPAFB, Ohio.

[69]   X-15 WSPO Weekly Activity Report, 23 January 1959, in the files of the AFMC History Office, WPAFB, Ohio.

[70]   X-15 WSPO Weekly Activity Report, 13 March 1959, in the files of the AFMC History Office, WPAFB, Ohio.

[71]   X-15 WSPO Weekly Activity Report, 1 May 1959, in the files of the AFMC History Office, WPAFB, Ohio.

[72]   Interview, Lieutenant R. L. Panton, X-15 WSPO Director of Systems Management, ARDC, 1 June 1959, by R. S. Houston, History Branch, WADC, in the files of the Air Force Museum archives.

[73]   James E. Love, "History and Development of the X-15 Research Aircraft," not dated, DFRC History Office, p. 20.

[74]   Walter C. Williams, "X-15 Concept Evolution" (a paper in the *Proceedings of the X-15 30th Anniversary Celebration*, Dryden Flight Research Facility, Edwards, California, 8 June 1989, NASA CP-3105), p. 14.

[75]   Lawrence P. Greene, "Summary of Pertinent Problems and Current Status of the X-15 Airplane" (a paper presented at the NACA Conference on the Progress of the X-15 Project, Langley Aeronautical Laboratory, 25-26 October 1956), p. 250.

[76]   Harrison A. Storms, "X-15 Hardware Design Challenges" (a paper in the *Proceedings of the X-15 30th Anniversary Celebration*, Dryden Flight Research Facility, Edwards, California, 8 June 1989, NASA CP-3105), p. 27.

[77]   The "N" designation indicated that the aircraft had undergone permanent modifications to a non-standard configuration.

[78]   James E. Love, "History and Development of the X-15 Research Aircraft," not dated, DFRC History Office, p. 11.

[79]   Richard P. Hallion, *On the Frontier: Flight Research at Dryden*, (Washington, DC.: NASA SP-4303, 1984), p. 101.

Chapter 3

The Flight Research Program

During the ten years of flight operations, five major aircraft were involved in the X-15 flight research program. The three X-15s were designated X-15-1 (56-6670), X-15-2 (56-6671), and X-15-3 (56-6672). Early in the test program the first two X-15s were essentially identical in configuration; the third aircraft was completed with different electronic and flight control systems. When the second aircraft was extensively modified after an accident midway through the test program, it became the X-15A-2. The two carrier aircraft were an NB-52A (52-003) and an NB-52B (52-008); they were essentially interchangeable.

The program used a three-part designation for each flight. The first number represented the specific X-15; 1 was for X-15-1, etc. No differentiation was made between the original X-15-2 and the modified X-15A-2. The second position was the flight number for that specific X-15. This included free-flights only, not captive-carries or aborts; the first flight was 1, the second 2, etc. If the flight was a scheduled captive-carry, the second position in the designation was replaced with a C; if it was an aborted free-flight attempt, it was replaced with an A. The third position was the total number of times any X-15 had been carried aloft by either NB-52. This

number incremented for each captive-carry, abort, and actual release. The 24 May 1960 letter from FRC Director Paul Bikle establishing this system is included as an appendix to this monograph.

Initial Flight Tests

The X-15-1 arrived at the Air Force Flight Test Center at Edwards AFB, California, on 17 October 1958; trucked over the hills from the North American plant in Los Angeles for testing at the NASA High-Speed Flight Station. It was joined by the second airplane in April 1959; the third would arrive later. In contrast to the relative secrecy that had attended flight tests with the XS-1 (X-1) a decade before, the X-15 program offered the spectacle of pure theater.[1]

As part of the X-15's contractor program, North American had to demonstrate each aircraft's general airworthiness during flights above Mach 2 before delivering it to the Air Force, which would then turn it over to NASA. Anything beyond Mach 3 was considered a part of the government's research obligation. The contractor program would last approximately two years, from mid-1959 through mid-1960.

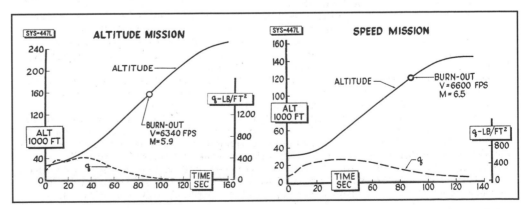

Two different mission profiles were flown— one for maximum speed; and one for maximum altitude. (NASA)

The first X-15 (56-6670) immediately prior to the official roll-out ceremonies at North American's Los Angeles plant on 15 October 1958. The small size of the trapezoid-shaped wings and the extreme wedge section of the vertical stabilizer are noteworthy. (North American Aviation)

The task of flying the X-15 during the contractor program rested in the capable hands of Scott Crossfield. After various ground checks, the X-15-1 was mated to the NB-52A, then more ground tests were conducted. On 10 March 1959, the pair made a scheduled captive-carry flight (program flight number 1-C-1). They had a gross take-off weight of 258,000 pounds, lifting off at 168 knots after a ground roll of 6,200 feet. During the 1 hour and 8 minute flight it was found that the NB-52 was an excellent carrier for the X-15, as had been expected from numerous wind tunnel and simulator tests. During the captive flight the X-15 flight controls were exercised and airspeed data from the flight test boom on the nose was obtained in order to calibrate the instrumentation. The penalties imposed by the X-15 on the NB-52 flight characteristics was found to be minimal in the gear-up configuration. The mated pair was flown up to Mach 0.83 at 45,000 feet.[2]

The next step was to release the X-15 from the NB-52 in order to ascertain its gliding and landing characteristics. The first glide flight was scheduled for 1 April 1959, but was aborted when the X-15 radio failed. The pair spent 1 hour and 45 minutes airborne conducting further tests in the mated configuration. A second attempt was aborted on 10 April 1959 by a combination of radio failure and APU problems. Yet a third attempt was aborted on 21 May 1959 when the X-15's stability augmentation system failed, and a bearing in the No. 1 APU overheated after approximately 29 minutes of operation.

The problems with the APU were the most disturbing. Various valve malfunctions, leaks, and several APU speed-control problems were encountered during these three flights, all of which would have been unacceptable during research flights. Tests conducted on the APU revealed that extremely high surge pressures were occurring at the pressure relief valve (actually a blow-out plug) during initial peroxide tank pressurization. The installation of an orifice in the helium pressurization line immediately downstream of the shut-off valve reduced the surges to acceptable levels. Other problems were found to be unique to the captive-carry flights and the long-run times being imposed on the APUs; they were deemed to be of little consequence to the flight test program

Long before the NB-52 first carried the X-15 into the air, engineers had tested the separation characteristics in the wind tunnels at Langley and Ames. Here an X-15 model drops-away from a model of the NB-52. Note that the X-15 is mounted on the wrong wing. This was necessary because the viewing area of the wind tunnel was on the left side of the aircraft. (NASA photo EL-1996-00114)

since the operating scenario would be different. The APUs underwent a constant set of minor improvements early in the flight test program, but nevertheless continued to be a source of irritation until the end.

On 22 May the first ground run of the interim XLR11 engine installation was accomplished using the X-15-2. Scott Crossfield was in the cockpit, and the test was considered successful, clearing the way for the eventual first powered flight; if the first X-15 could ever make its scheduled glide flight.

Another attempt at the glide flight was made on 5 June 1959 but was aborted even before the NB-52 left the ground[3] when Crossfield reported smoke in the X-15-1 cockpit. Investigation showed that a cockpit ventilation fan motor had overheated.

Finally, at 08:38 on 8 June 1959, Scott Crossfield separated the X-15-1 from the NB-52A at Mach 0.79 and 37,500 feet. Just prior to launch the pitch damper failed, but Crossfield elected to proceed with the flight, and switched the SAS pitch channel to standby. At launch, the X-15 separated cleanly and Crossfield rolled to the right with a bank

angle of about 30 degrees. The X-15 touched down on the dry lake at Edwards 4 minutes and 56 seconds later. Just prior to landing, the X-15 began a series of increasingly wild pitching motions; mostly as a result of Crossfield's instinctive corrective action, the airplane landed safely. Landing speed was 150 knots, and the X-15 rolled-out 3,900 feet while turning very slightly to the right. North American subsequently modified the control system boost to increase the control rate response, effectively solving the problem.

Although the impact at landing was not considered to be particularly hard, later inspection revealed that bell cranks in both main landing skids had bent slightly. The main skids were not instrumented on this flight, so no specific impact data could be ascertained, but it was generally believed that the shock struts had bottomed and remained bottomed as a result of higher than predicted landing loads. As a precaution against the main skid problem occurring again, the metering characteristics of the shock struts were improved, and lakebed drop tests at higher than previous loads were made with the landing gear test trailer that had been used to qualify the landing gear design. All other airplane sys-

North American test pilot A. Scott Crossfield was responsible for demonstrating that the X-15 was airworthy. His decision to leave NACA and join North American effectively locked him out of the high-speed and high-altitude test flights later in the program. (NASA photo EC-570-1

tems operated satisfactorily, clearing the way for the first powered flight.[4]

In preparation for the first powered flight, the X-15-2 was taken for a captive-carry flight with full propellant tanks on 24 July 1959. During August and early September, several attempts to make the first powered flight were cancelled before leaving the ground due to leaks in the APU peroxide system and hydraulic leaks. There were also several failures of propellant tank pressure regulators. Engineers and technicians worked to eliminate these problems, all of which were irritating, but none of which was considered critical.

The first powered flight was made by X-15-2 on 17 September 1959. The aircraft was released from the NB-52A at 08:08 in the morning while flying at Mach 0.80 and 37,600 feet. Crossfield piloted the X-15-2 to Mach 2.11 and 52,341 feet during 224.3 seconds of powered flight using the two XLR11 engines. He landed on the dry lakebed at Edwards 9 minutes and 11 seconds after launch. Following the landing, a fire was noticed in the area around the ventral stabilizer, and was quickly extinguished by ground

crews. A subsequent investigation revealed that the upper XLR11 fuel pump diffuser case had cracked after engine shutdown and had sprayed fuel throughout the engine compartment. Fuel collected in the ventral stabilizer and was ignited by unknown causes during landing. No appreciable damage was done, and the aircraft was quickly repaired.[5]

The third flight of X-15-2 took place on 5 November 1959 when the X-15 was dropped from the NB-52A at Mach 0.82 and 44,000 feet. During the engine start sequence, one chamber in the lower engine exploded. There was external damage around the engine and base plate, plus quite a bit of damage internal to the engine compartment. The resulting fire convinced Crossfield to make an emergency landing at Rosamond Dry Lake; he quickly shut off the engines, dumped the remaining fuel, and jettisoned the ventral[6] rudder. Even so, within the 13.9 seconds of powered flight, the X-15 managed to accelerate to Mach 1. The aircraft touched down near the center of the lake at approximately 150 knots and an 11 degrees angle of attack. When the nose gear bottomed out, the fuselage literally broke in half at station[7] 226.8, with about 70 percent of the bolts at the manufacturing

Any landing you can walk away from …

The X-15-2 made a hard landing on 5 November 1959, breaking its back as the nose settled on the lakebed. The damage looked worse than it was, and the aircraft was back in the air three months later. (NASA photo E-9543)

joint being sheared out. The fuselage contacted the ground and was dragged for approximately 1,500 feet. Crossfield later stated that the damage was the result of a defect that should have broken on the first flight.[8] The aircraft was sent to the North American plant for repairs, and was subsequently returned to Edwards in time for its fourth flight on 11 February 1960.[9]

The X-15-1 made its first powered flight, using two XLR11s, on 23 January 1960. This was also the first flight using the stable platform, and the performance of the system was considered encouraging. Under the terms of the contract, the X-15 had still "belonged" to North American until they had demonstrated its basic airworthiness and operation. Following this flight, a pre-delivery inspection was accomplished, and on 3 February 1960 the airplane was formally accepted by the Air Force and subsequently turned over to NASA.

The first government X-15 flight (1-3-8) was on 25 March 1960 with NASA test pilot Joseph A. Walker at the controls. The X-15-1 was launched at Mach 0.82 and 45,500 feet, although the stable platform had malfunctioned just prior to release. Two restarts were required on the top engine before all eight chambers were firing, and the flight lasted just over 9 minutes, reaching Mach 2.0 and 48,630 feet. For the next six months, Walker and Major Robert M. White alternated flying the X-15-1.[10]

It is interesting to note that the predictions regarding flutter made by Lawrence P. Greene at the first industry conference in 1956 did materialize, although fortunately they were not major and relatively easy to correct. During the early test flights, vibrations at 110 cycles had been noted and were the cause of some concern. Engineers at FRC added instrumentation to the X-15s from flight to flight in an attempt to isolate the vibrations and understand their origins, while wind tunnel tests were conducted at Langley. It was finally determined that the vibrations were

being caused by a flutter of the fuselage side tunnel panels. These had been constructed in removable sections with an unsupported length of over 6 feet in some cases.[11] North American added longitudinal stiffeners along the underside of each panel, and this largely cured the problem.[12]

The X-15-1 flew three times in the two weeks between 4 August and 19 August 1960, with five aborted launches due to various problems (including persistent APU failures). Two of these flights were made by Joe Walker, and one by Bob White. The flight on 12 August was to an altitude of 136,500 feet, marking the highest flight of an XLR11-powered X-15.

The Million Horsepower Engine[13]

The X-15-3 had arrived at Edwards on 29 June 1959 but had not yet flown when the first XLR99 flight engine (s/n 105) was installed in it during early 1960. It should be noted that the third X-15 was never equipped for the XLR11 engines. At the same time, the second X-15 was removed from flight status after its ninth flight (2-9-18) on 26 April 1960, in anticipation of replacing the XLR11 engines with the new XLR99. This left only the X-15-1 on active flight status.

The first ground run with the XLR99 in the X-15-3 was made on 2 June 1960. Inspection of the aircraft afterward revealed damage to the liquid oxygen inlet line brackets, the result of a water-hammer effect. After repairs were completed, another ground run was conducted on 8 June. A normal engine start and a short run at minimal power was made, followed by a normal shutdown. A restart was attempted, but was shutdown automatically by a malfunction indication. Almost immediately, a second restart was attempted, resulting in an explosion that effectively destroyed the aircraft aft of the wing. Crossfield was in the cockpit, which was thrown 30 feet forward, but he was not injured. Subsequent investigation revealed that the ammonia tank pressure regulator had failed open. Because of some ground han-

The top and bottom of the fuselage were usually covered in frost because the LOX tank was integral with the fuselage. Oxygen is liquid at -297 degrees Fahrenheit.

All three X-15s normally carried a yellow NASA banner on their vertical stabilizers. (U.S. Air Force)

dling hoses attached to the fuel vent line, the fuel pressure-relief valve did not operate properly, thus allowing the fuel tank to over-pressurize and rupture. In the process, the peroxide tank was damaged by debris, and the mixing of the peroxide and ammonia caused an explosion.

Post-accident analysis indicated that there were no serious design flaws with either the XLR99 or the X-15. The accident had been caused by a simple failure of the pressure regulator, exasperated by the unique configuration required for the ground test. Modification of the X-15-2 to accept the XLR99 continued, and several other modifications were incorporated at the same time. These included a revised vent system in the fuel tanks as an additional precaution against another explosion; revised ballistic control system components; and provisions for the installation of the ball-nose instead of the flight test boom that had been used so far in the program. The remains of the X-15-3 were returned to North American, which received authorization to rebuild the aircraft in early August.[14]

The installation of the ball-nose presented its own challenges since it had no capability to determine airspeed. The X-15 was designed with an alternate airspeed probe just forward of the cockpit, although two other locations, one well forward on the bottom centerline of the aircraft, and one somewhat aft near the centerline, had been considered alternate locations. Several early flights compared the data available from each location, while relying on the data provided by the airspeed sensors on the flight test boom protruding from the extreme nose. This indicated that the data from all three locations were acceptable, so the original location was retained. After the ball-nose was installed, angle-of-attack data was compared to that from previous flights using the flight test boom; the data were generally in good agreement, clearing the way for operational use of the ball-nose.

The first flight attempt of X-15-2 with the XLR99 was made on 13 October 1960, but was terminated prior to launch because of a peroxide leak in the No. 2 APU. Just to show how many things could go wrong on a single flight, there was also propellant impingement on the aft fuselage during the prime cycle, manifold pressure fluctuations during engine turbopump operation, and fuel tank pressure fluctuations during the jettison cycle. Nevertheless, two weeks later, Crossfield again entered the cockpit with the goal of making the first XLR99 flight. Again, problems with the No. 2 APU forced an abort.

On 15 November 1960, everything went right, and Crossfield made the first flight of X-15-2 powered by the XLR99. The primary flight objective was to demonstrate engine operation at 50 percent thrust. The launch was at 46,000 feet and Mach 0.83, and even with only half the available power, the X-15 managed to climb to 81,200 feet and Mach 2.97. The second XLR99 flight tested the engine's restart and throttling capability. Crossfield made the flight on 22 November, again using the second X-15. The third and final XLR99 demonstration flight was accomplished using X-15-2 on 6 December 1960. The objectives of engine throttling, shutdown, and restart were successfully accomplished. This marked North American Aviation's, and Scott Crossfield's, last X-15 flight. The job of flying the X-15 was now totally in the hands of the government test pilots.[15]

After this flight, a work schedule was established which would permit an early flight with a government pilot using North American maintenance personnel. The flight was tentatively scheduled for 21 December 1960 with Bob White as the pilot. However, a considerable amount of work had to be accomplished before the flight, including the removal and replacement of the engine (s/n 103) which had suffered excessive chamber coating loss, installation of redesigned canopy hooks, installation of an unrestricted upper vertical stabilizer, rearrangement of the alternate airspeed system, and the relocation of the ammonia tank helium pressure regulator into the fixed portion of the upper vertical. During a preflight ground run, a pinhole leak was found in the chamber throat of the engine. Although the leak was found to be acceptable for an engine run, it became increasingly worse during the test until it was such that the engine could not be run again. Since there was no spare engine available, the flight was cancelled and a schedule established to deliver the aircraft to the government prior to another flight. The X-15-2 was formally delivered to the Air Force and turned over to NASA on 8 February 1961. On the same day, X-15-1 was returned to the North American plant for conversion to the XLR99, having completed the last XLR11 flight of the program the day before with White at the controls.[16]

From the beginning of the X-15 flight test program in 1959 until the end of 1960, a total of 31 flights had been made with the first two

Six of the X-15 pilots (from left to right): Lieutenant Colonel Robert A. Rushworth (USAF), John B. "Jack" McKay (NASA), Lieutenant Commander Forrest S. Petersen (USN), Joseph A. Walker (NASA), Neil A. Armstrong (NASA), Major Robert M. White (USAF). (NASA via the San Diego Aerospace Museum Collection)

X-15s by seven pilots. But the X-15-1 was experiencing an odd problem. When the APU was started, hydraulic pressure was either slow in coming up, or dropped off out of limits when the control surfaces were moved. The solution to the problem came after additional instrumentation was placed on the hydraulic system. The boot-strap line which pressurized the hydraulic reservoir was freezing, causing a flow restriction or stoppage. Under these conditions, the hydraulic pump would cavitate, resulting in little or no pressure rise. The apparent cause of this problem was the addition of a liquid nitrogen line to cool the stable platform. Since the nitrogen line was installed adjacent to the hydraulic lines, it caused the Orinite hydraulic oil to freeze. The solution to the problem was to add electric heaters to the affected hydraulic lines.

Joe Walker's flight on 30 March 1961 marked the first use of the new A/P-22S full-pressure suit instead of the earlier MC-2. Walker reported the suit was much more comfortable and afforded better vision. But the flight pointed out a potential problem with the stability augmentation system (SAS). As Walker descended through 100,000 feet, a heavy vibration occurred and continued for about 45 seconds until recovery was affected at 55,000 feet. Incremental acceleration of approximately 1-g was noted in the vertical and transverse axes at a frequency of 13 cycles. This corresponded to the first bending mode of the horizontal stabilator. The center of gravity of the horizontal surfaces was located behind the hinge line; consequently rapid surface movement produced both rolling and pitching inertial moments. Subsequent analysis showed the vibration was sustained by the SAS at the natural frequency of the horizontal surfaces. Essentially, the oscillations began because of the increased activity of the controls on reentry which excited the oscillation and stopped after the pilot reduced the pitch-damper gain.[17]

Two solutions to the problem were discussed

between the FRC, North American, the Air Force, and the manufacturer of the SAS, Westinghouse; a notch filter for the SAS and a pressure-derivative feedback valve for the main stabilator hydraulic actuator. The notch filter eliminated SAS control surface input at 13 cycles, and the feedback valve damped the stabilator bending mode. In essence, the valve corrected the source of the problem, while the notch filter avoided the problem. Although it was felt that either solution would likely cure the problem, the final decision was to use both.

NASA research pilot William Dana made a check flight in a specially-modified JF-100C (53-1709) at Ames on 1 November 1960, delivering the aircraft to the FRC the following day. The aircraft had been modified as a variable-stability trainer that could simulate the X-15's flight profile. This made it possible to investigate new piloting techniques and control-law modifications without using an X-15. Another 104 flights were made for pilot checkout, variable stability research, and X-15 support before the aircraft was returned to Ames on 11 March 1964.[18]

The first government flight with the XLR99 engine took place on 7 March 1961 with Bob White at the controls. The X-15-2 reached Mach 4.43 and 77,450 feet, and the flight was generally satisfactory. The objectives of the flight were to obtain additional aerodynamic and structural heating data, as well as information on stability and control of the aircraft at high speeds. Post-flight examination showed a limited amount of buckling to the side-fuselage tunnels, attributed to thermal expansion. The temperature difference between the tunnel panels and the primary fuselage structure was close to 500 degrees Fahrenheit. The damage was not considered significant since the panels were not primary structure, but were only necessary to carry air loads. However, the buckling continued to become more severe as Mach numbers increased in later flights, and eventually NASA elected to install additional expansion joints in the tunnel skin to minimize the buckling.[19]

By June 1961, government test pilots had been operating the X-15 on research flights for just over a year.[20] The research phase of the X-15's flight program involved four broad objectives: verification of predicted hypersonic aerodynamic behavior and heating rates, study of the X-15's structural characteristics in an environment of high heating and high flight loads, investigation of hypersonic stability and control problems during atmospheric exit and reentry, and investigation of piloting tasks and pilot performance. By late 1961, these four areas had been generally examined, although detailed research continued to about 1964 using the first and third aircraft, and to 1967 with the second (as the X-15A-2). Before the end of 1961, the X-15 had attained its Mach 6 design goal and had flown well above 200,000 feet; by the end of 1962 the X-15 was routinely flying above 300,000 feet. The X-15 had already extended the range of winged aircraft flight speeds from Mach 3.2[21] to Mach 6.04, the latter achieved by Bob White on 9 November 1961.

The X-15 flight research program revealed a number of interesting things. Physiologists discovered the heart rates of X-15 pilots varied between 145 and 185 beats per minute in flight, as compared to a normal of 70 to 80 beats per minute for test missions in other aircraft. Researchers eventually concluded that pre-launch anticipatory stress, rather than actual post launch physical stress, influenced the heart rate. They believed, correctly, that these rates could be considered as probable baselines for predicting the physiological behavior of future astronauts. Aerodynamic researchers found remarkable agreement between the wind tunnel tests of exceedingly small X-15 models and actual results, with the exception of drag measurements. Drag produced by the blunt aft end of the actual aircraft proved 15 percent higher than wind tunnel tests had predicted.

At Mach 6, the X-15 absorbed eight times the heating load it experienced at Mach 3, with the highest heating rates occurring in the frontal and lower surfaces of the aircraft,

which received the brunt of airflow impact. During the first Mach 5+ flight, four expansion slots in the leading edge of the wing generated turbulent vortices that increased heating rates to the point that the external skin behind the joints buckled. It offered "… a classical example of the interaction among aerodynamic flow, thermodynamic properties of air, and elastic characteristics of structure." As a solution, small Inconel X alloy strips were added over the slots and additional fasteners on the skin.[22]

Heating and turbulent flow generated by the protruding cockpit enclosure posed other problems; on two occasions, the outer panels of the X-15's glass windshields fractured because heating loads in the expanding frame overstressed the soda-lime glass. The difficulty was overcome by changing the cockpit frame from Inconel X to titanium, eliminating the rear support (allowing the windscreen to expand slightly), and replacing the outer glass panels with high temperature alumina silica glass. All this warned aerospace designers to proceed cautiously. During 1968 John Becker[23] wrote: "The really important lesson here is that what are minor and unimportant features of a subsonic or supersonic aircraft must be dealt with as prime design problems in a hypersonic airplane. This lesson was applied effectively in the precise design of a host of important details on the manned space vehicles."

A serious roll instability predicted for the airplane under certain reentry conditions posed a dilemma to flight researchers. To accurately simulate the reentry profile of a returning winged spacecraft, the X-15 had to fly at angles of attack of at least 17 degrees. Yet the wedge-shaped vertical and ventral stabilizers, so necessary for stability and control in other portions of the flight regime, actually prevented the airplane from being flown safely at angles of attack greater than 20 degrees because of potential rolling problems. By this time, FRC researchers had gained enough experience with the XLR99 engine to realize that fears of thrust mis-

A common sight during the 1960s over Edwards—an NB-52 carrying an X-15. This was a boy's dream at the time; and the subject of many fantasies.

Over the course of the program, the markings on the NB-52s changed significantly. Early on, they were natural metal with bright orange verticals; later they were overall gray. (NASA)

alignment—a major reason for the large surfaces—were unwarranted. The obvious solution was simply to remove the lower portion of the ventral, something that X-15 pilots had to jettison prior to landing anyway so that the aircraft could touch down on its landing skids. Removing part of the ventral produced an acceptable tradeoff; while it reduced stability by about 50 percent at high angles of attack, it greatly improved the pilot's ability to control the airplane. With the ventral off, the X-15 could fly into the previously "uncontrollable" region above 20 degrees angle of attack with complete safety. Eventually the X-15 went on to reentry trajectories of up to 26 degrees, often with flight path angles of −38 degrees at speeds up to Mach 6.[24] Its reentry characteristics were remarkably similar to those of the later Space Shuttle orbiter.

When Project Mercury began, it rapidly eclipsed the X-15 in the public's imagination. It also dominated some of the research areas that had first interested X-15 planners, such as "zero-g" weightlessness studies. The use of reaction controls to maintain attitude in space proved academic after Mercury flew, but the X-15 would furnish valuable information on the blending of reaction controls with conventional aerodynamic con-

trols during exit and reentry, a matter of concern to subsequent Shuttle development. The X-15 experience clearly demonstrated the ability of pilots to fly rocket-propelled aircraft out of the atmosphere and back in to precision landings. Paul Bikle saw the X-15 and Mercury as a "… parallel, two-pronged approach to solving some of the problems of manned space flight. While Mercury was demonstrating man's capability to function effectively in space, the X-15 was demonstrating man's ability to control a high performance vehicle in a near-space environment … considerable new knowledge was obtained on the techniques and problems associated with lifting reentry."[25]

Nearly all of the early XLR99 flights experienced malfunction shutdowns of the engine immediately after launch, and sometimes after normal engine shutdown or burnout. Since the only active engine system after shutdown was the lube-oil system, investigations centered on it. Analyses of this condition revealed very wide acceleration excursions during the engine-start phase. A reasonable simulation of this acceleration was accomplished by placing an engine on a work stand with the ability to rotate the engine about the Y-axis. Under certain conditions, the lube-oil pump could be made to

cavitate for about 2 seconds, tripping an automatic malfunction shutdown. To eliminate this problem, a delay timer was installed in the lube-oil malfunction circuit which allowed the pump to cavitate up to 6 seconds without actuating the malfunction shutdown system. After this delay timer was installed in early 1962, no further engine shutdowns of this type were experienced.[26]

But a potentially more serious XLR99 problem was the unexpected loss of the Rokide coating from the combustion chamber during firing. A meeting was held at Wright Field on 13 June 1961 to discuss possible solutions. It was decided that the Wright Field Materials Laboratory would develop a new ceramic coating for the chambers, and that FRC would develop the technique and fixtures required to recoat chambers at Edwards. Originally, the Materials Laboratory awarded a contract to Plasmakote Corp. to perform the coating of several chambers, but the results were unsatisfactory. By March 1962, the techniques and fixtures developed by the FRC allowed chambers to be successfully recoated at Edwards.

Early in the program, the X-15's stability

augmentation and inertial guidance systems were two major problem areas. NASA eventually replaced the Sperry inertial unit with a Honeywell system designed for the stillborn Dyna-Soar. The propellant system had its own weaknesses; pneumatic vent and relief valves and pressure regulators gave the greatest difficulties, followed by spring pressure switches in the APUs, the turbopump, and the gas generation system. NASA's mechanics routinely had to reject 24-30 percent of spare parts as unusable, a clear indication of the difficulties that would be experienced later in the space programs in getting parts manufactured to exacting specifications.[27] Weather posed a critical factor. Many times Edwards enjoyed good weather while other locations on the High Range were covered with clouds, alternate landing sites were flooded, or some other meteorological condition postponed a mission.

Follow-on Experiments

During the summer of 1961, a new research initiative was proposed by the Air Force's Aeronautical Systems Division at Wright-Patterson AFB and NASA Headquarters: using the X-15 to carry a wide range of sci-

On 4 November 1960, the program attempted to launch two X-15 flights in a single day. Here X-15-1 is mounted on the NB-52B and X-15-2 is on the NB-52A. Rushworth was making his first flight in X-15-1, a low (48,900 feet) and slow (Mach 1.95) familiarization. The X-15-2, with Crossfield as pilot, aborted due to a failure in the No. 2 APU. (NASA photo E-6186)

entific experiments unforeseen when the aircraft was conceived in 1954.

Researchers at the FRC wanted to use the X-15 to carry high-altitude experiments related to the proposed Orbiting Astronomical Observatory; others suggested modifying one of the airplanes to carry a Mach 5+ ramjet for advanced air-breathing propulsion studies. Over 40 experiments were suggested by the scientific community as suitable candidates for the X-15 to carry. In August 1961 NASA and the Air Force formed the "X-15 Joint Program Coordinating Committee" to prepare a plan for a follow-on experiments program. The committee held its first meeting on 23-25 August 1961 at the FRC.[28]

Many experiments suggested to the committee related to space science, such as ultraviolet stellar photography. Others supported the Apollo program and hypersonic ramjet studies. Hartley Soulé and John Stack, then NASA's director of aeronautical research, proposed the classification of experiments into two groups: category A experiments consisted of well-advanced and funded experiments having great importance; category B included worthwhile projects of less urgency or importance.[29]

In March 1962 the committee approved the "X-15 Follow-on Program," and NASA announced that an ultraviolet stellar photography experiment from the University of Wisconsin's Washburn Observatory would be first. The X-15's space science program eventually included twenty-eight experiments including astronomy, micrometeorite collection (using wing-top pods on the X-15-1 and X-15-3 that opened at 150,000 feet), and high-altitude mapping. The micrometeorite experiment was unsuccessful, and was ultimately cancelled. Two of the follow-on programs, a horizon definition experiment from the Massachusetts Institute of Technology, and test of insulation material for the Saturn launch vehicle, directly benefited the Apollo program. The Saturn insulation was applied to the X-15's speed brakes,

which were then deployed at the desired speed and dynamic pressure to test both the insulating properties and the bonding material. By the end of 1964, over 65 percent of data being returned from the three X-15 aircraft involved follow-on projects; this percentage increased yearly through conclusion of the program.[30]

As early as May 1962, North American had proposed modifying one of the X-15s as a flying test bed for hypersonic engines. Since the X-15s were being fully utilized at the time, neither the Air Force nor NASA expressed much interest in pursuing the idea. However, when the X-15-2 was damaged during a landing accident on 9 November 1962 (seriously injuring Jack McKay, who would later return from his injuries to fly the X-15 again), North American proposed modifying the aircraft in conjunction with its repairs. General support for the plan was found within the Air Force, which was willing to pay the estimated $6 million.[31]

On the other hand, NASA was less enthusiastic, and felt the aircraft should simply be repaired to its original configuration.[32] Researchers at NASA believed that the Mach 8 X-15 would prove to be of limited value for propulsion research. However, NASA did not press its views, and in March 1963 the Air Force authorized North American to rebuild the aircraft as the X-15A-2. Twenty-nine inches were added to the fuselage between the existing propellant tanks. The extra volume was to be used by a liquid hydrogen tank to power the ramjet, but the LH2 tank could be replaced by other equipment as needed. In fact, the compartment was frequently used to house cameras to test reconnaissance concepts, or to observe the dummy ramjet during flight tests, through three heat-resistant windows in the lower fuselage. The capability to carry two external propellant tanks was added to provide additional powered flight with the XLR99. The right wingtip was also modified to allow various wingtip shapes to be carried interchangeably, although it appears that this capability was never used.[33]

Forty weeks and $9 million later, North American delivered the X-15A-2.[34] The aircraft made its first flight on 25 June 1964 piloted by Bob Rushworth. Early flights demonstrated that the aircraft retained satisfactory flying qualities at Mach 5, although on three flights thermal stresses caused portions of the landing gear to extend at Mach 4.3, generating "an awful bang and a yaw."[35] In each case Rushworth landed safely, despite the blow-out of the heat-weakened tires in one instance. On 18 November 1966, Pete Knight set an unofficial world's speed record of Mach 6.33 in the aircraft. The drop tanks had been jettisoned at Mach 2.27 and 69,700 feet. A nonfunctional dummy ramjet was constructed in order to gather aerodynamic data on the basic shape in preparation for possible flight tests in the early 1970s. The first flight with the dummy ramjet attached to the ventral was on 8 May 1967. Although providing a pronounced nose-down trim change, the ramjet actually restored some of the directional stability lost when the lower ventral rudder had been removed.

NASA had evaluated several possible coatings that could be applied over the X-15's Inconel X hot-structure to enable it to withstand the thermal loads experienced above Mach 6. The use of such coatings could be beneficial since various ablators were being investigated by the major aerospace contractors during the early pre-concept phases[36] of the Space Shuttle development.[37] Such a coating would have to be relatively light, have good insulating properties, and be easy to apply, remove, and reapply before another flight. The selected coating was MA-25S, an ablator developed by the Martin Company in connection with some early reusable spacecraft studies. Consisting of a resin base, a catalyst, and a glass bead powder, it would protect the hot-structure from the expected 2,000 degrees Fahrenheit heating at Mach 8. Martin estimated that the coating, ranging from 0.59 inches thick on the canopy, wings, vertical, and horizontal stabilizers, down to 0.015 inches on the trailing edges of the wings and tail, would keep the skin temperature below 600 degrees Fahrenheit. The first unpleasant surprise came, however, with the application of the coating to the X-15A-2: it took six weeks. Getting the correct thickness over the entire surface proved harder than expected. Also, every time a panel had to be opened to service the X-15, the coating had to be removed and reapplied around the affected area.

Because the ablator would char and emit a residue in flight, North American had installed an "eyelid" over the left cockpit window; it would remain closed until just before landing. During launch and climbout, the pilot would use the right window, but residue from the ablator would render it opaque above Mach 6. The eyelid had already been tested on several flights.[38]

Late in the summer of 1967, the X-15A-2 was ready for flight with the ablative coating. The weight of the ablator—125 pounds higher than planned—together with expected increased drag reduced the theoretical maximum performance of the airplane to Mach 7.4, still a significant advance over the Mach 6.3 previously attained. The appearance of the X-15A-2 was striking, an overall flat off-white finish, the external tanks a mix of silver and orange-red with broad striping. On 21 August 1967, Knight completed the first flight in the ablative coated X-15A-2, reaching Mach 4.94 and familiarizing himself with its handling qualities. His next flight was destined to be the program's fastest flight, and the last flight of the X-15A-2.[39]

On 3 October 1967, 43,750 feet over Mud Lake, Knight dropped away from the NB-52B. The flight plan showed the X-15A-2 would weigh 52,117 pounds at separation, more than 50 percent heavier than originally conceived in 1954.[40] The external tanks were jettisoned 67.4 seconds after launch at Mach 2.4 and 72,300 feet; tank separation was satisfactory, however, Knight felt the ejection was "harder" than the last one he had experienced (2-50-89). The recovery system performed satisfactorily and the tanks were recovered in repairable condition. The XLR99 burned for

140.7 seconds before Knight shut it down. Radar data showed the X-15A-2 attained Mach 6.70 (4,520 mph) at 102,700 feet, a winged-vehicle speed record that would stand until the return of the Space Shuttle *Columbia* from its first orbital flight in 1981.[41]

The post-landing inspection revealed many things. The ability of the ablative material to protect the aircraft structure from the high aerodynamic heating was considered good except in the area around the dummy ramjet where the heating rates were significantly higher than predicted. The instrumentation on the dummy ramjet had ceased working approximately 25 seconds after engine shut-down, indicating that a burn through of the ramjet/pylon structure had occurred. Shortly thereafter the heat propagated upward into the lower aft fuselage causing the hydrogen-peroxide hot light to illuminate in the cock-pit. Assuming a genuine overheat condition, William Dana in the NASA 1 control room had requested Knight to jettison the remaining peroxide. The high heat in the aft fuse-lage area also caused a failure of a helium check valve allowing not only the normal helium source gas to escape, but also the emergency jettison control gas supply as well. Thus, the remaining residual propellants could not be jettisoned. The aircraft was an estimated 1,500 pounds heavier than normal at landing, but the landing occurred without incident.

Engineers had not fully considered possible shock interaction with the ramjet shape at hypersonic speeds. As it turned out, the flow patterns were such that a tremendous shock wave impinged on the ramjet and its sup-porting structure. The heat in the ramjet pylon area was later estimated to be ten times normal, and became high enough at some time during the flight to ignite 3 of the 4 explosive bolts holding the ramjet to the pylon. As Knight was turning downwind in the landing pattern, the one remaining bolt failed structurally and the ramjet separated from the aircraft. Knight did not feel the ramjet separate, and since the chase aircraft had not yet joined up, was unaware that the ramjet had separated.

The position of the X-15 at the time of sepa-ration was later established by radar data and the most likely trajectory estimated. A ground search party discovered the ramjet on the Edwards bombing range. Although it had been damaged by impact, it was returned for study of the heat damage.

The unprotected right-hand windshield was, as anticipated, partially covered with ablation products. Since the left eyelid remained closed until well into the recovery maneuver, Knight flew the X-15 using on-board instru-ments and directions from William Dana in the NASA 1 control room. The eyelid was opened at approximately Mach 1.6 as the air-craft was over Rogers Dry Lake, and the visi-bility was considered satisfactory. Knight landed at Edwards 8 minutes and 12 seconds after launch.

The ablator obviously was not totally success-ful; in fact this was the closest any X-15 came to structural failure induced by heating. Post-flight inspection revealed that the aircraft was

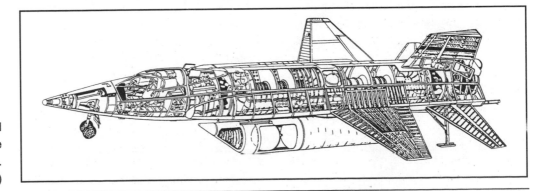

An internal general arrangement of the modified X-15A-2. (NASA)

charred on its leading edges and nose. The ablator had actually prevented cooling of some hot spots by keeping the heat away from the hot-structure. Some heating effects, such as where shock waves impinged on the ramjet had not been thoroughly studied. To John Becker the flight underscored "… the need for maximum attention to aerothermodynamic detail in design and preflight testing."[42] To Jack Kolf, an X-15 project engineer at the FRC, the post-flight condition of the airplane "… was a surprise to all of us. If there had been any question that the airplane was going to come back in that shape, we never would have flown it."[43]

Some of the problems encountered with the ablator were nonrepresentative of possible future uses. The X-15 had been designed as an uninsulated hot structure. Any future vehicle would probably be designed with a more conventional airframe, eliminating some of the problems encountered on this flight. But some of the problems were very

real. The amount of time it took to apply the ablator was unacceptable. Even considering that the learning curve was steep, and that after some experience the time could be cut in half or even further, the six weeks it took to coat the relatively small X-15 bode ill for larger vehicles. Nevertheless, ablators would continue to be proposed on various Space Shuttle concepts, in decreasing quantity, until 1970 when several forms of ceramic tiles and metal "shingles" would become the preferred concepts.[44]

It was estimated that repairing the X-15A-2 and refurbishing the ablator for another flight near Mach 7 would have taken five weeks. The unexpected airflow problems around the ramjet ended any idea of flying it again. NASA sent the X-15A-2 to North American for general maintenance and repair, and although the aircraft returned to Edwards in June 1968, it never flew again. It is now on exhibit—in natural black finish—at the Air Force Museum, Wright-Patterson AFB, Ohio.

The X-15A-2 drops away from the NB-52 on its last flight. Note the dummy ramjet attached to the ventral and the overall white finish applied to the ablator. The drop tanks would be jettisoned 67.4 seconds after engine ignition, at a speed of Mach 2.4 and 72,300 feet altitude. Pete Knight would attain Mach 6.70 on this flight. (NASA)

Ultimately, Garrett did deliver a functioning model of the ramjet, and it was successfully tested in a wind tunnel in late 1969. In this case successful meant that supersonic combustion was achieved, although for a very short duration and under very controlled and controversial conditions.[45]

Adaptive Controls

The X-15-3 featured specialized flight instrumentation and displays that rendered it particularly suitable for high-altitude flight research. A key element was the Minneapolis Honeywell MH-96 "adaptive" flight control system originally developed for the X-20 Dyna-Soar. This system automatically compensated for the airplane's behavior in various flight regimes, combining the aerodynamic control surfaces and the reaction controls into a single control package. This was obviously the way future high-speed aircraft and spacecraft would be controlled, but the technology of the 1960s were severely taxed by the requirements for such a system.

By the end of 1963, the X-15-3 had flown above 50 miles altitude. This was the altitude that the Air Force recognized as the minimum boundary of space flight, and five Air Force pilots were awarded Astronaut Wings for their flights in the X-15.[46] All but one of these flights was with X-15-3 (Astronaut Joe Engle's first space flight was in X-15-1). NASA did not recognize the 50 mile criteria, using the international 62 mile standard instead. Only a single NASA pilot went this high; Joe Walker set a record for winged space flight by reaching 354,200 feet (67 miles), a record that stood until the orbital flight of *Columbia* nearly two decades later. By mid-1967, the X-15-3 had completed sixty-four research flights, twenty-one at altitudes above 200,000 feet. It became the primary aircraft for carrying experiments to high altitude.

The X-15-3 would also make the most tragic flight of the program. At 10:30 in the morning on 15 November 1967, the X-15-3 dropped away from the NB-52B at 45,000 feet over Delamar Dry Lake. At the controls was Major Michael J. Adams, making his seventh X-15 flight. Starting his climb under full power, he was soon passing through 85,000 feet. Then an electrical disturbance distracted him and slightly degraded the control of the aircraft; having adequate backup controls, Adams continued on. At 10:33 he reached a peak altitude of 266,000 feet. In the NASA 1 control room, mission controller Pete Knight monitored the mission with a team of engineers. As the X-15 climbed, Adams started a planned wing-rocking maneuver so an on-board camera could scan the horizon. The wing rocking quickly became excessive, by a factor of two or three. At the conclusion of the wing-rocking portion of the climb, the X-15 began a slow drift in heading; 40 seconds later, when the aircraft reached its maximum altitude, it was off heading by 15 degrees. As Adams came over the top, the drift briefly halted, with the airplane yawed 15 degrees to the right. Then the drift began again; within 30 seconds, Adams was descending at right angles to the flight path. At 230,000 feet, encountering rapidly increasing dynamic pressures, the X-15 entered a Mach 5 spin.[47]

In the NASA 1 control room there was no way to monitor heading, so nobody suspected the true situation that Adams now faced. The controllers did not know that the airplane was yawing, eventually turning completely around. In fact, Knight advised Adams that he was "a little bit high," but in "real good shape." Just 15 seconds later, Adams radioed that the aircraft "seems squirrely." At 10:34 came a shattering call: "I'm in a spin, Pete." Plagued by lack of heading information, the control room staff saw only large and very slow pitching and rolling motions. One reaction was "disbelief; the feeling that possibly he was overstating the case." But Adams again called out, "I'm in a spin." As best they could, the ground controllers sought to get the X-15 straightened out. There was no recommended spin recovery technique for the X-15, and engineers knew nothing about the aircraft's

realizing that the X-15 would never make Rogers Dry Lake, went into afterburner and raced for the emergency lakes; Ballarat and Cuddeback. Adams held the X-15's controls against the spin, using both the aerodynamic control surfaces and the reaction controls. Through some combination of pilot technique and basic aerodynamic stability, the airplane recovered from the spin at 118,000 feet and went into an inverted Mach 4.7 dive at an angle between 40 and 45 degrees.[48]

Adams was in a relatively high altitude dive and had a good chance of rolling upright, pulling out, and setting up a landing. But now came a technical problem; the MH-96 began a limit-cycle oscillation just as the airplane came out of the spin, preventing the gain changer from reducing pitch as dynamic pressure increased. The X-15 began a rapid pitching motion of increasing severity, still in a dive at 160,000 feet per minute, dynamic pressure increasing intolerably. As the X-15 neared 65,000 feet, it was diving at Mach 3.93 and experiencing over 15-g vertically, both positive and negative, and 8-g laterally.

The aircraft broke up northeast of the town of Johannesburg 10 minutes and 35 seconds after launch. A chase pilot spotted dust on Cuddeback, but it was not the X-15. Then an Air Force pilot, who had been up on a delayed chase mission and had tagged along on the X-15 flight to see if he could fill in for an errant chase plane, spotted the main wreckage northwest of Cuddeback. Mike Adams was dead; the X-15-3 destroyed.[49]

NASA and the Air Force convened an accident board. Chaired by NASA's Donald R. Bellman, the board took two months to prepare its report. Ground parties scoured the countryside looking for wreckage; critical to the investigation was the film from the cockpit camera. The weekend after the accident, an unofficial FRC search party found the camera; disappointingly, the film cartridge was nowhere in sight. Engineers theorized that the film cassette, being lighter than the camera, might be further away, blown north by winds at altitude. FRC engineer Victor Horton organized a search and on 29 November, during the first pass over the area, Willard E. Dives found the cassette.

Most puzzling was Adams' complete lack of awareness of major heading deviations in spite of accurately functioning cockpit instrumentation. The accident board concluded that he had allowed the aircraft to deviate as the result of a combination of distraction, misinterpretation of his instrumentation display, and possible vertigo. The electrical disturbance early in the flight degraded the overall effectiveness of the aircraft's control system and further added to pilot workload. The MH-96 adaptive control system then caused the airplane to break up during reentry. The board made two major recommendations: install a telemetered heading indicator in the control room, visible to the flight controller; and medically screen X-15 pilot candidates for labyrinth (vertigo) sensitivity.[50] As a result of the X-15's crash, the FRC added a ground-based "8 ball" attitude indicator in the control room to furnish mission controllers with real time pitch, roll, heading, angle of attack, and sideslip information.

Mike Adams was posthumously awarded Astronaut Wings for his last flight in the X-15-3, which had attained an altitude of 266,000 feet—50.38 miles. In 1991 Adams' name was added to the Astronaut Memorial at the Kennedy Space Center in Florida.

The X-15 program would only fly another eight missions. The X-15A-2, grounded for repairs, soon remained grounded forever. The X-15-1 continued flying, with sharp differences of opinion about whether the research results returned were worth the risk and expense.

A proposed delta wing modification to the X-15-3 had offered supporters the hope that the program might continue to 1972 or 1973. The delta wing X-15 had grown out of studies in the early 1960s on using the X-15 as a hypersonic cruise research vehicle.

Essentially, the delta wing X-15 would have made use of the third airframe with the adaptive flight control system, but also incorporated the modifications made to the X-15A-2—lengthening the fuselage, revising the landing gear, adding external propellant tanks, and provisions for a small-scale experimental ramjet. NASA proponents, particularly John Becker at Langley, found the idea very attractive since: "The highly swept delta wing has emerged from studies of the past decade as the form most likely to be utilized on future hypersonic flight vehicles in which high lift/drag ratio is a prime requirement i.e., hypersonic transports and military hypersonic cruise vehicles, and certain recoverable boost vehicles as well."[51]

Despite such endorsement, support remained lukewarm at best both within NASA and the Air Force; the loss of Mike Adams and the X-15-3 effectively ended all thought of such a modification.

As early as March 1964, in consultation with NASA Headquarters, Brigadier General James T. Stewart, director of science and technology for the Air Force, had determined to end the X-15 program by 1968.[52] At a meeting of the Aeronautics/Astronautics Coordinating Board on 5 July 1966, it was decided that NASA should assume total responsibility for all X-15 costs (other than incidental AFFTC support) on 1 January 1968.[53] This was later postponed one year. As it turned out, by December 1968 only the X-15-1 was still flying, and it cost roughly $600,000 per flight. Other NASA programs could benefit from this funding, and thus NASA did not request a continuation of X-15 funding after December 1968.[54] During 1968 William Dana and Pete Knight took turns flying the X-15-1. On 24 October 1968, Dana completed the X-15's 199th, and as it turned out the last, flight reaching Mach 5.38 at 255,000 feet. A total of ten attempts were made to launch the 200th flight, but a variety of maintenance and weather problems forced cancellation every time. On 20 December 1968, the X-15-1 was demated from the NB-52A for the last time. After nearly a decade of flight operations, the X-15 program came to an end.

The instrument panel of the X-15-3 with the MH-96 adaptive control system installed. The dark panel immediately ahead of the center control stick allowed the pilot to control how the MH-96 reacted. (NASA photo E63-9834)

[1] Richard P. Hallion, editor, *The Hypersonic Revolution: Case Studies in the History of Hypersonic Technology* (Aeronautical Systems Division, Wright-Patterson AFB, Ohio, 1987), Volume I, *Transiting from Air to Space: The North American X-15*, p. 129.

[2] James E. Love, "History and Development of the X-15 Research Aircraft," not dated, DFRC History Office, p. 12.

[3] Because the NB-52 never left the ground, this attempt does not have a program flight number.

[4] James E. Love, "History and Development of the X-15 Research Aircraft," not dated, DFRC History Office, p. 13.

[5] Ibid.

[6] The configuration of the X-15 included a short fixed vertical stabilizer with an all-moving rudder above it on top of the fuselage, and short fixed ventral stabilizer with a jettisonable all-moving rudder below it. This lower rudder will be called the ventral rudder for simplicity and clarity.

[7] Fuselage stations are measured in inches from a fixed point on the nose of the aircraft.

[8] A. Scott Crossfield, during an interview in the NBC documentary film *The Rocket Pilots*, 1989.

[9] James E. Love, "History and Development of the X-15 Research Aircraft," not dated, DFRC History Office, p. 14.

[10] Ibid.

[11] Concerns over panel flutter resulted in extensive redesign of the proposed X-20 Dyna-Soar.

[12] James E. Love, "History and Development of the X-15 Research Aircraft," not dated, DFRC History Office, p. 14.

[13] The XLR99 was popularly considered to be a million horsepower engine. By the definition in Websters, the horsepower of a rocket engine is determined by multiplying the thrust (in pounds) times the speed (in mph), divided by 375. Therefore, the XLR99 would be 57,000 lbf * 4,520 mph / 375 = 687,040 hp. Not quite a million, but still impressive for an 800 pound engine.

[14] James E. Love, "History and Development of the X-15 Research Aircraft," not dated, DFRC History Office, p. 17.

[15] Ibid., pp. 15-16.

[16] *X-15 Research Airplane Flight Record*, North American Aviation, NA-65-1, revised 15 May 1968.

[17] James E. Love, "History and Development of the X-15 Research Aircraft," not dated, DFRC History Office, p. 21.

[18] Email from Peter W. Merlin, DFRC History Office, 18 November 1999.

[19] James E. Love, "History and Development of the X-15 Research Aircraft," not dated, DFRC History Office, p. 20.

[20] A. Scott Crossfield and Clay Blair, *Always Another Dawn: The Story of a Rocket Test Pilot*, World Publishing Co., 1960, pp. 307-366.

[21] Achieved by Captain Milburn Apt in the first Bell X-2 on 27 September 1956. Unfortunately, Apt was killed on this flight.

[22] Wendell H. Stillwell, *X-15 Research Results*, (Scientific and Technical Information Branch, NASA, Washington, DC.: NASA SP-60, 1965), pp. 65.

[23] John V. Becker, "The X-15 Program in Retrospect" (paper presented at the 3rd Eugen Sänger Memorial Lecture, Bonn, Germany, 4-5 December 1968); and James E. Love, *X-15: Past and Future*, paper presented to the Fort Wayne Section, Society of Automotive Engineers, 9 December 1964.

[24] Wendell H. Stillwell, *X-15 Research Results*, (Scientific and Technical Information Branch, NASA, Washington, DC.: NASA SP-60, 1965), pp. 51-52.

[25] Wendell H. Stillwell, *X-15 Research Results*, (Scientific and Technical Information Branch, NASA, Washington, DC.: NASA SP-60, 1965). p. iv; see also Walter C. Williams, "The Role of the Pilot in the Mercury and X-15 Flights" (in the *Proceedings of the Fourteenth AGARD General Assembly*, 16-17 September 1965, Portugal).

[26] James E. Love, "History and Development of the X-15 Research Aircraft," not dated, DFRC History Office, p. 17.

[27] James E. Love, and William R. Young, *Survey of Operation and Cost Experience of the X-15 Airplane as a Reusable Space Vehicle* (Washington, DC: NASA TN-D-3732, 1966).

[28] James E. Love, "History and Development of the X-15 Research Aircraft," not dated, DFRC History Office, p. 23.

[29] Memorandum from Homer Newell to Hugh L. Dryden, 18 December 1961, subject: X-15 follow-on program; Memorandum from Paul F. Bikle to Hartley Soulé (probably November 1961); NASA news release 61-261. All in the files of the NASA Dryden History Office.

[30] Air Force Systems Command, X-15 System Package Program, 6-37-48; NASA news release 62-98; X-15 news release 62-91; Letter from Hugh L. Dryden to Lieutenant General James Ferguson, 15 July 1963; NASA news release 64-42. All in the files of the NASA Dryden History Office.

[31] John V. Becker, "A Hindsight Study of the NASA Hypersonic Rocket Engine Program" (Washington DC., NASA/OAST, 1 July 1976), p. 9, in the files of the NASA History Office.

[32] Memorandum from Paul F. Bikle to the NASA Office of Aeronautical Research, subject: Repairs to the X-15-2 Airplane, 27 December 1962, in the files of the NASA History Office.

[33] Ronald G. Boston, "Outline of the X-15's Contributions to Aerospace Technology," typescript available in the NASA Dryden History Office, p. 12.

[34] Robert A. Hoover and Robert A. Rushworth, "X-15A-2 Advanced Capability" (a paper presented at the annual symposium of The Society of Experimental Test Pilots, Beverly Hills, California, 25-26 September 1964).

[35] Cockpit voice transcription for Rushworth flight, 14 August 1964; X-15 Operations Flight Report, 19 August 1964; Rushworth flight comments, not dated. All in the files of the NASA Dryden History Office.

[36] These included contracts from both the Air Force and various NASA centers as part of the Integral Launch and Reentry Vehicle (ILRV) programs.

[37] Dennis R. Jenkins, *The History of Developing the National Space Transportation System: The Beginning through STS-75* (second edition; Cape Canaveral, Florida: Dennis R. Jenkins, 1997), p. 129.

[38] Robert A. Hoover and Robert A. Rushworth, "X-15A-2 Advanced Capability" (a paper presented at the annual symposium of The Society of Experimental Test Pilots, Beverly Hills, California, 25-26 September 1964).

[39] William J. Knight, "Increased Piloting Tasks and Performance of X-15A-2 in Hypersonic Flight" (a paper presented at the annual symposium of the SETP, Beverly Hills, CA, 28–30 September 1967).

[40] Flight Plan for Flight Number 2-53-97, 19 September 1967.

[41] James R. Welsh, "Preliminary Report on X-15 Flight 2-53-97" (X-15 Planning Office, 26 October 1967).

[42] John V. Becker, "The X-15 Program in Retrospect" (paper presented at the 3rd Eugen Sänger Memorial Lecture, Bonn, Germany, 4-5 December 1968).

[43] Interview with Jack Koll, 28 February 1977, (interviewer unknown), in the files of the AFSC History Office.

[44] Dennis R. Jenkins, *The History of Developing the National Space Transportation System: The Beginning Through STS-75* (second edition; Cape Canaveral, Florida: Dennis R. Jenkins, 1997), p. 108.

[45] Ronald G. Boston, "Outline of the X-15's Contributions to Aerospace Technology," typescript available in the NASA Dryden History Office, p. 8.

[46] Major Michael J. Adams, Captain Joseph H. Engle, Major William J. "Pete" Knight, Lieutenant Colonel Robert A. Rushworth, and Major Robert M. White.

[47] To date, this is the only hypersonic spin that has been encountered during manned flight research. See Bellman, Donald R., et al., *Investigation of the Crash of the X-15-3 Aircraft on November 15, 1967*, January 1968, pp. 8-15.

[48] Ibid.

[49] Ibid.

[50] Ibid.

[51] Memorandum from John V. Becker to Floyd Thompson, 29 October 1964; Letter from Paul F. Bikle to C. W. Harper, 13 November 1964.

[52] USAF Headquarters Development Directive No. 32, 5 March 1964, reprinted in X-15 System Package Program, 13-7.

[53] Becker, John V. and Supp, R. E., *Report of Meeting of USAF/NASA Working Groups on Hypersonic Aircraft Technology*, 21-22 September 1966.

[54] Love, James E. and Young, William R., NASA TN D-3732, *Survey of Operation and Cost Experience of the X-15 Airplane as a Reusable Space Vehicle*, November 1966, pp. 7.

Jack McKay was seriously injured on Flight 2-31-52, 9 November 1962. The XLR99 stuck at 35 percent thrust, forcing McKay to abort. The flaps did not extend fully, resulting in a fast landing on Mud Lake. The aircraft rolled over after touchdown. McKay recovered and came back to fly the X-15 22 more times. (NASA photo E-9149)

The X-15A-2 being prepared for Flight 2-43-75 on 3 November 1965. This was the first flight with the external propellant tanks, which were empty. The tanks were painted bright orange and white to aid in photography during separation. (NASA)

All three X-15s are lined up in the main hangar at the Flight Research Center in 1966. Note the lifting bodies in the background, along with an F-4A, F5D, and DC-3. (NASA photo EC66-1461)

Six of the twelve men to fly X-15 pose for a portrait in 1966 Left to right): Captain Joseph H. Engle (USAF), Major Robert A. Rushworth (USAF), John B. "Jack" McKay (NASA), William J. "Pete" Knight (NASA), Milton O. Thompson (NASA), and William H. Dana (NASA). (NASA Photo EC66-1017)

Chapter 4

The Legacy of the X-15

The year 1999 marked the 40th anniversary of the first flight of the X-15; this anniversary occurred more than 30 years after the program ended. The X-15 was the last high-speed research aircraft to fly as part of the research airplane program. The stillborn X-30 of the 1980s never took flight, and the verdict is still out on the fate of the Lockheed Martin X-33 demonstrator. Neil Armstrong, among others, once called the X-15 "the most successful research airplane in history."[1]

Twelve men flew X-15. Scott Crossfield was first; William Dana was last. Pete Knight went 4,520 mph (Mach 6.70); Joe Walker went 67 miles (354,200 feet) high. Five of the pilots were awarded Astronaut Wings. Mike Adams died. What was learned? What should have been learned?

In October 1968 John V. Becker enumerated 22 accomplishments from the research and development work that produced the X-15, 28 accomplishments from its actual flight research, and 16 from experiments carried by the X-15. Becker's comments have been well documented elsewhere, but are quoted here as appropriate.[2]

Nearly ten years after Becker's assessment, Captain Ronald G. Boston of the U.S. Air Force Academy's history department reviewed the X-15 program for "lessons learned" that might benefit the development of the X-24C National Hypersonic Flight Research Facility Program, an effort that was cancelled shortly afterwards. Boston's paper offered an interesting perspective on the X-15 from the vantage point of the mid-1970s.[3]

In 1999, the historian at the Dryden Flight Research Center, J. D. "Dill" Hunley, wrote a lessons-learned paper on the X-15. Drawing heavily but not uncritically upon Becker's and Boston's insights, it too provides an interesting perspective, and is quoted several times in the pages that follow.[4]

Lessons Learned (or not)

The X-15 was designed to achieve a speed of Mach 6 and an altitude of 250,000 feet to explore the hypersonic and near-space environments. More specifically, its goals were:

(1) to verify existing (1954) theory and wind tunnel techniques;

(2) to study aircraft structures under high (1,200 degrees Fahrenheit) heating;

(3) to investigate stability and control problems associated with high-altitude boost and reentry; and

(4) to investigate the biomedical effects of both weightless and high-g flight.

All of these design goals were met, and most were surpassed. The X-15 actually achieved Mach 6.70, 354,200 feet, 1,350 degrees Fahrenheit, and dynamic pressures over 2,200 pounds per square foot.[5] In addition, once the original research goals were achieved, the X-15 became a high-altitude hypersonic testbed for which 46 follow-on experiments were designed.

Unfortunately due to the absence of a subsequent hypersonic mission, aircraft applications of X-15 technology have been few. Given the major advances in materials and computer technology in the 30 years since the end of the flight research program, it is unlikely that many of the actual hardware lessons are still applicable. That being said, the lessons learned from hypersonic modeling, simulation, and the insight gained by being able to evaluate actual X-15 flight test results against wind tunnel and predicted results, greatly expanded the confidence of researchers during the 1960s and 1970s.

In space, however, the X-15 contributed significantly to both the Apollo and Space Shuttle programs. Perhaps the major contribution was the final elimination of a spray-on ablator as a possible thermal protection system for the Space Shuttle. This would likely have happened in any case as the ceramic tiles and metal shingles were further developed, but the operational problems encountered with the (admittedly brief) experience on X-15A-2 hastened the departure of the ablators. Although largely intangible, proving the value of man-in-the-loop simulations and precision "dead-stick" landings have also been invaluable to the Space Shuttle program.

The full value of X-15's experience to designing advanced aircraft and spacecraft can only be guessed at. Many of the engineers (including Harrison Storms) from the X-15 project worked on the Apollo spacecraft and the Space Shuttle. In fact, the X-15 experience may have been part of the reason that North American was selected to build later spacecraft. Yet X-15's experience is overshadowed by more recent projects and becomes difficult to trace as systems evolve through successive programs. Nonetheless, many of those engineers are confident that they owe much to the X-15, even if many are at a loss to give any concrete examples.

Political Considerations

John V. Becker, arguably the father of the X-15, once stated that the project came along at " ... the most propitious of all possible times for its promotion and approval." At the time it was not considered necessary to have a defined operational program in order to conduct basic research. There were no

"glamorous and expensive" manned space projects to compete for funding, and the general feeling within the nation was one of trying to go faster, higher, or further. In today's environment, as in 1968 when Becker was commenting, it is highly unlikely that a program such as the X-15 could gain approval.[6]

This situation should give pause to those who fund aerospace projects solely on the basis of their presumably predictable outcomes and their expected cost effectiveness. Without the X-15's pioneering work, it is quite possible that the manned space program would have been slowed, conceivably with disastrous consequences for national prestige.[7]

According to Becker, proceeding with a general research configuration rather than with a prototype of a vehicle designed to achieve a specific mission as envisioned in 1954 was critical to the ultimate success the X-15 enjoyed. Had the prototype route been taken, Becker believed that " ... we would have picked the wrong mission, the wrong structure, the wrong aerodynamic shapes, and the wrong propulsion." He also believed that a second vital aspect to the success of the X-15 was its ability to conduct research, albeit for very short periods of time, outside the sensible atmosphere.[8]

The latter proved to be the most important aspect of X-15 research, given the contributions it made to the space program. But in 1954 this could not have been foreseen. Few people then believed that flight into space was imminent, and most thought that flying humans into space was improbable before the next century. Fortunately, the hypersonic aspects of the proposed X-15 enjoyed "virtually unanimous approval," although ironically the space-oriented results of the X-15 have been of greater value than its contributions to aeronautics.[9]

A final lesson from the X-15 program is that success comes at a cost. It is highly likely that researchers can never accurately predict the costs of exploring the unknown. If you understand the problems well enough to accurately predict the cost, the research is not necessary. The original cost estimate for the X-15 program was $10.7 million. Actual costs were still a bargain in comparison with those for Apollo, Space Shuttle, and the International Space Station, but at $300 million, they were over almost 30 times the original estimate.[10] Because the X-15's costs were not subjected to the same scrutiny from the Administration and Congress that today's aerospace projects undergo, the program continued. One of the risks when exploring the unknown is that you do not understand all the risks. Perhaps politicians and administrators should learn this particular lesson from this early and highly successful program.

Rocket Engines

The XLR99 was the first large man-rated rocket engine that was capable of being throttled and restarted in flight. This complexity resulted in many aborted missions (approximately one-tenth of all mission aborts) and significantly added to the development cost of the engine. When the X-15 program ended, many felt that the throttleable feature might have been a needless luxury that complicated and delayed the development of the XLR99.

But in the mid-1960s these attributes were considered vital to the development of a rocket engine to power the Space Shuttle. At the time, Shuttle was to consist of two totally reusable stages—essentially a large hypersonic aircraft that carried a smaller winged spacecraft much like the NB-52s carried the X-15s. The same basic engine was going to power both stages; the pilots therefore needed to be able to control its thrust output. At some points in the early Shuttle concept development phases, the same engines would also be used on-orbit to effect changes in the orbital plane. So the original concept for the Space Shuttle Main Engines (SSME) included the ability to operate at 10 percent of their rated thrust, and to be restarted multiple times during flight.[11]

In the end, the production SSMEs are throttleable within much the same range as the XLR99—65 to 109 percent, in one percent increments. In actuality about the only routine use of this ability is to throttle down as the vehicle reaches the point of maximum dynamic pressure during ascent, easing stresses on the vehicle for a few seconds on each flight. Even this would not have been necessary with a different design for the solid rocket boosters.[12] So the complexities required to enable the engine to throttle may, again, have been a needless luxury. Nevertheless, the development pains experienced by Reaction Motors provided insight for Pratt & Whitney and Rocketdyne (the two main SSME competitors) during the design and development of the SSMEs.

Human Factors

Coming at a time when serious doubts were being raised concerning man's ability to handle complex tasks in the high-speed, weightless environment of space, the X-15 became the first program for repetitive, dynamic monitoring of pilot heart rate, respiration, and EKG under extreme stress over a wide range of speeds and forces. The Bioastronautics Branch of the AFFTC measured unusually high heart and breathing rates on the parts of the X-15 pilots at points such as launch of the X-15 from the NB-52, engine shutdown, pullout from reentry, and landing. Heart rates averaged 145 to 160 beats per minute with peaks on some flights of up to 185 beats per minute. Despite the high levels, which caused initial concern, these heart rates were not associated with any physical problems or loss of ability to perform piloting tasks requiring considerable precision. Consequently, theoretical limits had to be re-evaluated, and Project Mercury as well as later space programs did not have to be concerned about such high heart rates in the absence of other symptoms. In fact, the X-15's data provided some of the confidence to go ahead with early manned Mercury flights—the downrange ballistic shots being not entirely dissimilar to the X-15's mission profile.[13]

The bio-instrumentation developed for the X-15 program has allowed similar monitoring of many subsequent flight test programs. Incorporated into the pressure suit, pickups are unencumbering and compatible with aircraft electronics. The flexible, spray-on wire leads have since found use in monitoring cardiac patients in ambulances.

Another contribution of the X-15 program was the development of what John Becker calls the "first practical full-pressure suit for pilot protection in space."[14] The David Clark Company had worked with the Navy and the HSFS on an early full-pressure suit for use in high-altitude flights of the Douglas D-558-II; the suit worn by Marion Carl on his high-altitude flights was the first step. This suit was made of a waffle-weave material and had only a cloth enclosure rather than a helmet. It should be noted that Scott Crossfield was heavily involved in the creation of this suit, the success of which Crossfield attributes to "... David Clark's genius."[15]

The David Clark Company later developed the A/P-22S-2 pressure suit that permitted a higher degree of mobility.[16] It consisted of a link-net material covering a rubberized pressure garment. Developed specifically for the X-15, the basic pressure suit provided part of the technological basis for the suits used in the Mercury and Gemini programs. It was later refined as the A/P-22S-6 suit that became the standard Air Force operational suit for high altitude flight in aircraft such as the U-2 and SR-71. However, it should be added that the space suit for Project Mercury underwent further development and was produced by the B.F. Goodrich Company rather than the David Clark Company, so the line of development from X-15 to Mercury was not entirely a linear one, and security surrounding the U-2 and Blackbird programs have obscured some of this history.[17]

X-15 pilots practiced in a ground-based simulator that included the X-15 cockpit with all of its switches, controls, gauges, and instruments. An analog computer converted the pilot's movements with the controls into instrument readings and indicated what the aircraft would do in flight to respond to control actions. After a flight planner had used the simulator to lay out a flight plan, the pilot and flight planner worked "for days and weeks practicing for a particular flight." The X-15 simulator was continually updated with data from previous flights to make it more accurate, and eventually a digital computer allowed it to perform at higher fidelity.[18]

Much has been made of the side-stick controller used on the X-15. Although the concept has found its way onto other aircraft, it has usually been for reasons other than those that initially drove its use on the X-15. The X-15 designers feared that the high g-loads encountered during acceleration would make it impossible for the pilot to use the conventional center stick; such worries are not the reason Airbus Industries has used the controller on the A318-series airliners. And although the side-stick controller has proven very popular in the F-16 fighter, it has not been widely adopted. Nevertheless, the X-15 experience provided a wealth of data over a wide range of flight regimes.

Some phases of X-15 flight, such as reentry, were marginally stable, and the aircraft required artificial augmentation (damping) systems to achieve satisfactory stability. The X-15 necessitated the development of an early stability augmentation system (SAS). The first two X-15s were equipped with a simple fail-safe, fixed-gain system. The X-15-3 was equipped with a triple-redundant adaptive flight control system; the pilot flew via inputs to the augmentation system. Although a point of continuing debate, the X-15 did not incorporate a "fly-by-wire" system if meant to denote a nonmechanically linked control system. Nevertheless, the SAS system did "fly" the X-15-3 based on pilot input rather than the pilot flying it directly. This basic concept would find use on an entire generation of aircraft, including such high performance fighters as the F-15. The advent of true fly-by-wire aircraft, such as the F/A-18, would advance the concept even further.

Aeronautics

In 1954, the few existing hypersonic wind tunnels were small and presumably unable to simulate the conditions of actual flight at speeds above Mach 5. The realistic fear at the time was that testing in them would fail to produce valid data. The X-15 provided the earliest, and so far most significant, validation of hypersonic wind tunnel data. This was of particular significance since it would be extremely difficult and very expensive to build a large-scale hypersonic wind tunnel.

This general validation, although broadly confirmed by other missiles and spacecraft, came primarily from the X-15; it made the conventional, low-temperature, hypersonic wind tunnel an accepted source of data for configuration development of hypersonic vehicles.[19]

The X-15 program offered an excellent opportunity to compare actual flight data with theory and wind tunnel predictions. The X-15 verified existing wind tunnel techniques for approximating interference effects for high-Mach, high angle-of-attack hypersonic flight, thus giving increased confidence in small-scale techniques for hypersonic design studies. Wind tunnel drag measurements were also validated, except for a 15 percent discrepancy found in base drag—caused by the sting support used in the wind tunnel. All of this greatly increased the confidence of engineers as they set about designing the Space Shuttle.

One of the widely held beliefs in the mid-1950s was the theoretical presumption that the boundary layer (the thin layer of air close to the surface of an aircraft) would be highly stable at hypersonic speeds because of heat flow away from it. This presumption fostered the belief that hypersonic aircraft would enjoy laminar (smooth) airflow over their surfaces. At Mach 6, even wind tunnel extrapolations indicated extensive laminar flow. However, flight data from the X-15 showed that only the leading edges exhibited laminar flow and that turbulent flow occurred over most surfaces. Small surface irregularities, which produced turbulent flow at transonic and supersonic speeds, also did so at Mach 6.[20] Thus, engineers had to abandon their hopeful expectations. Importantly, X-15 flight test data indicated that hypersonic flow phenomena were linear above Mach 5, allowing increased confidence during design of the Space Shuttle, which must routinely transition through Mach 25 on its way to and from space. The basic X-15 data were also very useful to the NASP designers while that program was viable.

In a major discovery, the Sommer-Short and Eckert T-prime aerodynamic heating prediction theories in use during the late 1950s were found to be 30 to 40 percent in excess of flight test results. Most specialists in fluid mechanics refused to believe the data, but repeated in-flight measurements completely substantiated the initial findings. This led the aerodynamicists to undertake renewed ground-based research to complete their understanding of the phenomena involved—highlighting the value of flight research in doing what Hugh Dryden had predicted for the X-15 in 1956: that it would "separate the real from the imagined."[21]

Subsequent wind tunnel testing led to Langley's adopting the empirical Spaulding-Chi model for hypersonic heating. This eventually allowed the design of lighter vehicles with less thermal protection that could more easily be launched into space. The Spaulding-Chi model found its first major use during the design of the Apollo command and service modules and proved to be quite accurate. In 1999 the Spaulding-Chi model was still the primary tool in use.

Based on their X-15 experience, North American devised a computerized mathematical model for aerodynamic heating called HASTE (Hypersonic and Supersonic Thermal Evaluation) which gave a workable "first cut" approximation for design studies. HASTE was, for example, used directly in the initial Apollo design study. Subsequent versions of this basic model were also used early in the Space Shuttle design evolution.

At the time of the first Mach 5 X-15 flight, perhaps its greatest contribution to aeronautics was to disprove the existence of a "stability barrier" to hypersonic flight that was suspected after earlier research aircraft encountered extreme instability at high supersonic speeds. Although of little consequence today, the development of the "wedge" tail allowed the X-15 to successfully fly above Mach 5 without the instability that had plagued the X-1 series and X-2 aircraft at much lower speeds. The advent of modern fly-by-wire controls and stability augmentation systems based around high speed digital computers have allowed designers to compensate for gross instabilities in basic aerodynamic design, and even to tailor an aircraft's behavior differently for different flight regimes. The era of building a vehicle that is dynamically stable has passed, and with it much of this lesson.

The art of simulation grew with the X-15 program, not only for pilot training and mission rehearsal, but for research into controllability problems. The same fixed-based simulator used by the pilots could also be used to explore those areas of the flight envelope deemed too risky for actual flight. The X-15 program showed the value of combining wind tunnel testing and simulation in maximizing the knowledge gained from each of the 199 test flights. It also provided a means of comparing "real" flight data with wind tunnel data. It is interesting to note that the man-in-the-loop simulation first used on X-15 found wide application on the X-30 and the X-33. In fact, DFRC research pilot Stephen D. Ishmael has flown hundreds of hours "in" the X-33, which ironically is an unpiloted vehicle.

Flight Research and Space Flight

Before the X-15, high-speed research aircraft flown at Edwards could be monitored and tracked from Edwards. The trajectory of the X-15 extended much farther from Edwards than those of the previous research aircraft, requiring two up-range stations where tracking, communications, and telemetry equipment were installed and integrated with the control room back at the FRC. Along the X-15 flight route, program personnel also surveyed a series of dry lakebeds for emergency landings and tested them before each flight to ensure they were hard enough to permit the X-15 to land.[22] In many ways this parallels the tracking and communications network and the transatlantic abort sites used by the Space Shuttle.

The opportunity to observe pilot performance under high stress and high g-forces indicated that an extensive ground training program was needed to prepare pilots to handle the complex tasks and mission profiles of space flight. The result was a simulation program that became the foundation for crew training for all human space flight. The program depended on four types of simulation.

Prior to the first X-15 mission, the ability of the pilot to function under the high g-forces expected during launch and reentry was tested in a closed-loop, six-degree-of-freedom centrifuge at the Naval Air Development Center, Johnsville, Pennsylvania. This project became the prototype for programs set up at the Ames Research Center and the Manned Spacecraft Center at Houston (now the Johnson Space Center).[23]

A static cockpit mockup provided the means for extensive mission rehearsal—averaging 20 hours per 10 minute flight. Such preparation was directly responsible for the high degree of mission success achieved as pilots rehearsed their primary, alternate, and emergency mission profiles. Similar, but much more elaborate, rehearsals are still used by astronauts preparing for Space Shuttle flights.

X-15 pilots maintained proficiency by flying an NT-33 or JF-100C variable-stability aircraft whose handling charac-

A great deal of what was learned on X-15 went on to build Space Shuttle. (NASA)

teristics could be varied in flight, simulating the varied response of the X-15 traversing a wide range of velocities and atmospheric densities. Much of this training is now conducted in advanced motion-based simulators, although the Air Force still operates a variable-stability aircraft (the VISTA F-16).

Pilots practiced the approach and landing maneuver in F-104 aircraft. With landing gear and speed brakes extended, the F-104's power-off glide ratio approximated that of the unpowered X-15. Shuttle crews continue this same practice using modified Gulfstream Shuttle Training Aircraft (STA).

Astronaut "capsule communicators," (capcomms) were a direct outgrowth of the X-15's practice of using an experienced pilot as the ground communicator for most X-15 missions.[24] This practice existed through Mercury, Gemini, and Apollo, and continues today on Space Shuttle missions. It is still believed that a pilot on the ground makes the best person to communicate with the crew, especially in stressful or emergency situations.[25]

Subsequent flight test work at Edwards relied heavily on the methodology developed

for the X-15. There are no fewer than three high-tech control facilities located at Edwards today; the facility at Dryden, the Riddley Control Center complex at the AFFTC, and the B-2 control complex located on South Base. Each of these control centers has multiple control rooms for use during flight test. The X-33 program has built yet another control room, this one located near the launch site at Haystack Butte.[26]

The X-15 program required a tracking network known as "High Range." Operational techniques were established for real-time flight monitoring which were carried over to the space program. The experience of setting up this control network became something of a legacy to Mercury and later space projects through the personnel involved. Gerald M. Truszynski, as Chief of the Instrumentation Division at the FRC, had participated in setting up the High Range, as had Edmond C. Buckley, who headed the Instrument Research Division at Langley. The Tracking and Ground Instrumentation Group at Langley had the responsibility for tracking the Mercury capsules, and it was headed, briefly, by Buckley.[27]

Buckley soon transferred to NASA Headquarters as assistant director for space flight operations, with Truszynski joining him in 1960 as an operations engineer. Both continued to be involved in instrumentation and communication until a reorganization under NASA Administrator James Webb created an Office of Tracking and Data Acquisition with Buckley as director. Buckley named Truszynski as his deputy, and in 1962 appointed him to lead the Apollo Task Group that shaped the Apollo tracking and data network.[28] Much of this same infrastructure was used early in the Space Shuttle program.

Meanwhile, Walter Williams, who had headed the NACA operations at the HSFS/FRC since 1946, was reassigned as Associate Director of the newly formed Space Task Group at Langley in September 1959. He eventually served as the Director of Operations for Mercury, and then as Associate Director of the Manned Spacecraft Center. He also served as operations director in the Mercury Control Center at Cape Canaveral during the Mercury flights of Alan Shepard, Gus Grissom, and John Glenn in 1961 and 1962.[29]

Experience from the NASA 1 control room undoubtedly influenced the development of the Mercury Control Center at Cape Canaveral, and perhaps more distantly, even the Mission Control Center (MCC)[30] at Houston.[31] However, the spacecraft control rooms and their tracking and data acquisition systems drew on many other sources (including the missile ranges which they shared),[32] although the experience setting up the High Range and operating the NASA 1 control room undoubtedly provided some operational perspectives.

An often overlooked area where the X-15 influenced Space Shuttle operations is in the energy management maneuvers immediately prior to landing. By demonstrating that it was possible to make precision unpowered landings with vehicles having a low lift-over-drag ratio, the X-15 program smoothed the path for the slightly later lifting-body program and then for the space shuttle procedures for energy management and landing.

The techniques used by X-15 pilots consisted of arriving at a "high key" above the intended landing point. Once he reached the high key, the pilot did not usually need or receive additional information from the control room; he could complete the landing using visual information and his own experience with practice landings in an F-104 configured to simulate an X-15 landing. With considerable variation on different missions, the pilot would arrive at the high key on an altitude mission at about 35,000 feet, turn 180 degrees and proceed to a "low key" at about 18,000 feet, where he would turn another 180 degrees and proceed to a landing on Rogers Dry Lake. Depending upon the amount of energy remaining, the pilot could use shallow or tight bank angles and speed brakes as necessary.

Because of their much higher energy, the standard approach for the Space Shuttle consists of a variation on this 360-degree approach. As a Shuttle approaches the runway for landing, if it has excess energy for a normal approach and landing, it dissipates this energy in S-turns (banking turns) until it can slow to a subsonic velocity at about 49,000 feet of altitude some 25 miles from the runway. It then begins the approach and landing phase at about 10,000 feet and an equivalent airspeed of about 320 mph some 8 miles from the runway.[33] Early in the Space Shuttle program, a specially-configured T-38[34] would accompany the orbiter on the final approach, much as the X-15 chase aircraft did at Edwards. Shuttle pilots practice in a specially-configured Gulfstream Shuttle Training Aircraft, much as the X-15 pilots did in the modified F-104.

Components and Construction

The X-15 was designed with a hot-structure that could absorb the heat generated by its short-duration flight. Remember, the X-15 seldom flew for over ten minutes at a time, and a much shorter time was spent at the maximum speed or dynamic pressure. Development showed the validity of ground "partial simulation" testing of primary members stressed under high temperature. A facility was later built at DFRC for heat-stress testing of the entire structure, and similar testing was accomplished on the YF-12A Blackbird and the Space Shuttle structural test article (STA-099).[35]

The X-15 pioneered the use of corrugations and beading to relieve thermal expansion stresses. Metals with dissimilar expansion coefficients were also used to alleviate stresses, and the leading edges were segmented to allow for expansion. Around the same time, similar techniques were apparently developed independently by Lockheed for use on Blackbird series of Mach 3+ aircraft.

The X-15 represented the first large-scale use of Inconel X, in addition to extensive use of titanium alloys. This required the development of new techniques for forming, milling, drilling, and welding that came to be widely used in the aerospace industry. North American pioneered chemical milling, a construction technique that has since been used on other projects.

The differentially deflected horizontal stabilizers on the X-15 provided roll and pitch control and allowed designers to eliminate the ailerons that would have provided a severe structural and theromodynamic problem within the thin wing section used on the X-15. This configuration was already being flight tested by less exotic aircraft (YF-107A) at the same time it was used on the X-15, but nevertheless proved extremely valuable. It is common practice today to use differential stabilators on modern aircraft, particularly fighters, although in most cases conventional ailerons are also retained; the flight control system deciding when to use which control surfaces based on conditions.

The all-moving vertical surfaces in lieu of conventional rudders has proven somewhat less attractive to aircraft designers. North American used an all-moving vertical surface on the A-5 Vigilante, designed not long after the X-15. Lockheed also used all-moving surfaces on the Blackbird series of Mach 3 aircraft, although it is difficult to ascertain if the X-15 influenced this design choice.

The X-15 designers also had to solve problems relating to high aerodynamic heating in proximity to cryogenic liquids. This led to cryogenic tubing that was used on parts of the Apollo spacecraft, and thermal insulation design features that were later used on the Space Shuttle. An early experience of running a liquid nitrogen cooling line too close to a hydraulic line taught designers about the need to fully understand the nature of the fluids they were dealing with. In-flight failures on high altitude flights with the X-15 also taught aerospace engineers about such things as the need to pressurize gear boxes on auxiliary power units to prevent foaming of the lubricant in the low pressure of space, since that led to material failures.[36]

Although the primary structure of the X-15 proved sound, several detailed design problems were uncovered during early flight tests. A surprise lesson came with the discovery of heretofore unconsidered local heating phenomena. Small slots in the wing leading edge, the abrupt contour change along the canopy, and the wing root caused flow disruptions that produced excessive heating and adjacent material failure. The X-15, tested in "typical" panels or sections, demonstrated the problems encountered when those sections are joined and thus precipitated an analytical program designed to predict such local heating stresses. From this experience, Rockwell engineers closely scrutinized the segmented carbon-carbon composite leading edge of the Space Shuttle wing. The bimetallic "floating retainer" concept designed to dissipate stresses across the X-15's windshield carried over to the Apollo and Space Shuttle windshield designs as well.

On three occasions, excessive aerodynamic heating of the nose-wheel door scoop caused structural deformation, permitting hot boundary-layer air to flow into the wheel well, damaging the landing gear, and in one case causing the gear to extend at Mach 4.2 (flight 2-33-56). Although the landing gear remained intact, the disintegration of the tires made the landing very rough. The need for very careful examination of all seals became apparent, and closer scrutiny of surface irregularities, small cracks, and areas of flow interaction became routine. The lessons learned from this influenced the final detailed design of the Space Shuttle to ensure that gaps and panel lines were adequately protected against inadvertent airflow entry.

Other problems from aerodynamic heating included windshield crazing, panel flutter, and skin buckling. Arguably, designers could have prevented these problems through more extensive ground testing and analysis, but a key purpose of flight research is to discover the unexpected. The truly significant lesson from these problems is that defect in subsonic or supersonic aircraft that are comparatively minor at slower speeds become much more critical at hypersonic speeds.[37]

One of the primary concerns during the X-15 development was panel flutter, evidenced by the closing paper presented at the 1956 industry conference. Panel flutter has proven difficult to predict at each speed increment throughout history, and the hypersonic regime was no different. Although the X-15 was conservatively designed, and incorporated all the lessons from first generation supersonic aircraft, the fuselage side tunnels and the vertical surfaces were prone to develop panel flutter during flight. This led to an industry-wide reevaluation of panel flutter design criteria in 1961-62. Stiffeners and reduced panel sizes alleviated the problems on the X-15, and similar techniques later found general application in the high speed aircraft of the 1960s.[38] The lessons learned at Mach 6 defined criteria later used in the development of the Space Shuttle.

The X-15 provided the first opportunity to study the effects of acoustical fatigue over a wide range of Mach numbers and dynamic

pressures. In these first in-flight measurements, "boundary layer noise"-related stresses were found to be a function of g-force, not Mach number. Such fatigue was determined to be no great problem for a structure stressed to normal in-flight loading. This knowledge has allowed for more optimum structural design of missiles and space capsules that experience high velocities.

On the X-15, the measurement of velocity was handled by early inertial systems. All three X-15s were initially equipped with analog-type systems which proved to be highly unreliable. Later, two aircraft, including the X-15-3 with the adaptive control system, were modified with digital systems. In the subsequent parallel evaluation of analog versus digital inertial systems, the latter was found to be far superior. It was far more flexible and could make direct inputs to the adaptive flight control system; it was also subject to less error. Thanks to advances in technology such as laser-ring gyros and digital computers, inertial systems have become inexpensive, highly accurate, and very reliable.[39] In recent years they have been integrated with the Global Positioning System (GPS), providing three-dimensional attitude and position information.

During the early test flights, the X-15 relied on simple pilot-static pressure instruments mounted on a typical flight test nose boom. These were not capable of functioning as speeds and altitudes increased. To provide attitude information, the NACA developed the null-sensing "ball-nose" which could survive the thermal environment of the X-15. An extendable pitot tube was added when the velocity envelope was expanded beyond Mach 6. Thus far the ball-nose has not found subsequent application, and probably never will since inertial and GPS systems have evolved so quickly. Interestingly, the Space Shuttle still uses an extendable pitot probe during the landing phase.

The X-15 was the first vehicle to routinely use reaction controls. The HSFS had begun research on reaction controls in the mid-1950s using a fixed-base analog control stick with a pilot presentation to determine the effects of control inputs. This was followed by a mechanical simulator to enable the pilot to experience the motions created by reaction controls. This device emulated the inertial ratios of the X-1B, which incorporated a reaction control system using hydrogen-peroxide as a monopropellant, decomposed by passing it through a silver-screen catalyst. Because of fatigue cracks later found in the fuel tank of the X-1B, it completed only three flights using the reaction control system before it was retired in 1958.[40]

As a result, a JF-104A with a somewhat more refined reaction control system was tested beginning in late 1959 and extending into 1961. The JF-104A flew a zoom-climb maneuver to achieve low dynamic pressures at about 80,000 feet that simulated those at higher altitudes. The techniques for using reaction controls on the X-15, and more importantly, for transferring from aerodynamic controls to reaction controls and back to aerodynamic controls provided a legacy to the space program.[41]

The X-15-3 was equipped with a Minneapolis Honeywell MH-96 self-adaptive control system designed for the cancelled Dyna-Soar. The other two X-15s had one controller on the right-hand side of the cockpit for aerodynamic controls and another on the left-hand side for the reaction controls. Thus, the pilot had to use both hands for control during the transition from flying in the atmosphere to flying outside the atmosphere and then back in the opposite direction. Since there was no static stability outside the atmosphere, the pilot had to counter any induced aircraft motion manually using the reaction controls. The MH-96 had an attitude hold feature that maintained the desired attitude except during control inputs. The MH-96 also integrated the aerodynamic and reaction controls in a single controller, greatly improving handling qualities during the transition from aerodynamic to space flight, as well as reducing pilot workload.[42]

But the basic feature of the MH-96 was automatic adjustment of gain (sensitivity) to maintain a desirable dynamic response of the airplane. The MH-96 compared the actual response of the airplane with a preconceived ideal response in terms of yaw, pitch, and roll rates. Initially, Milt Thompson stated that the system was "somewhat unnerving to the pilot" because he was not in "direct control of the aircraft" but was only "commanding a computer that then responded with its own idea of what is necessary in terms of a control output." However, pilots became "enthusiastic in their acceptance of it" when they realized that the MH-96 resulted in "more precise command than was possible" with the reaction controls by themselves. Consequently, the X-15-3 with the MH-96 was used for all altitude flights planned above 270,000 feet.[43]

There were some problems with the experimental system, including one that contributed to the death of Mike Adams in X-15-3 on 15 November 1967. Nevertheless, the MH-96 constituted a significant advance in technology that helped pave the way toward fly-by-wire in the early 1970s. Today, most every aircraft, and several automobiles, feature some variation of a fly-by-wire system with automatic rate-gain adjustment and stability augmentation.[44]

Follow-on Experiments

During the early 1960s, only the X-15 had the capability to carry a useful payload above the atmosphere. And unlike missiles, the X-15 could return equipment and results for reevaluation, recalibration, and reuse. Perhaps the earliest true "follow-on" experiment came in 1961: a coating material designed to reduce the infrared emissions of the proposed B-70 was tested to Mach 4.43 (525 degrees Fahrenheit) on the exterior surface of an X-15 stabilizer panel. Thus began a series of 46 follow-on experiments in physical sciences, space navigation aids, reconnaissance studies, and advanced aerodynamics. While not all of the 46 experiments were completed before the X-15 program ended, many of them did yield useful data.

Heating: Throughout the X-15's flight career it participated in heating studies, mainly to verify the output from wind tunnels and computer simulations. Late in the flight program, one X-15 was fitted with a sharp leading edge on the upper vertical stabilizer, and the results were compared with theory and with data from the original blunt leading edge.

Astronomy: The ultraviolet stellar photography study measured the ultraviolet brightness of several stars to determine their material composition. The X-15 carried four cameras on a gimbaled platform in the instrument bay behind the cockpit above the filtering effects of the ozone layer—approximately 40 miles altitude. Conducted in 1963 and again in 1966, this work was subsequently continued on improved sounding rockets, then on orbital satellites.

The X-15 was ideally suited to measure atmospheric densities at altitudes of 50,000 to 235,000 feet, cross-checking measurements on ascent with those on descent. Using the ball-nose to take measurements, flow-angularity errors were eliminated. The X-15 provided atmospheric seasonal variation density profiles. Unfortunately, these measurements could only be taken in the area immediately around Edwards AFB.

The X-15 provided the first direct solar spectrum measurement of the Sun from above the atmosphere. A scientific revelation, this data allowed the refinement of the Solar Constant of Radiation which was revalued 2.5 percent lower than existing ground-based determinations. This constant provides a measure of thermal energy incident on the Earth and upon which all photochemical processes depend. It is also useful for the design of thermal protection for spacecraft.[45]

Micrometeorites: Designed to collect micrometeorites at various altitudes, this experiment was part of a larger NASA study to build a particle-impact data base for spacecraft design criteria. Only on the last of six flights did this experiment "catch" any particles, and those were so contaminated by the exhaust from the reaction controls that the project was cancelled.

Space Navigation: The X-15 supported two—MIT and NASA-Langley—horizon definition projects to determine the Earth's infrared horizon radiance profile. This information was later used in attitude referencing systems for orbiting spacecraft. The MIT work was part of an Apollo support program seeking alternative means for orbit reinsertion guidance in case of radar or communications failure. The space sextant designed for this task was checked enroute on Apollo missions 8, 10, and 11 with relatively good accuracy when compared to radar position.

A successful program to collect data on radiation characteristics of the daytime sky background was part of an effort to develop a "star tracking" navigational system. Star trackers went on to be used aboard U-2 and SR-71 aircraft, and two of them are mounted in the forward fuselage of each Space Shuttle orbiter.[46]

Reconnaissance Systems: The X-15's performance made it an ideal testbed for high-speed aircraft and satellite reconnaissance systems. Ultraviolet (UV) sensors were studied as a means of detecting incoming ICBMs. This three-part project yielded promising results, but UV systems were overshadowed by the more advanced infrared systems. In an effort to determine the exhaust plume signature of a typical rocket exhaust above the ozone layer, the exhaust plume of the X-15 itself was scanned. To test the feasibility of detecting a missile launch by its UV signature, an actual launch from Vandenberg AFB was scheduled to be monitored on X-15 flight number 200, but this never occurred.

Several infrared (IR) satellite detection systems began as X-15 experiments. As early as 1963, researchers studied the IR exhaust plume characteristics of the X-15. A follow-up project to measure the Earth's infrared background using an IR scanner never flew before the X-15 program ended. Nonetheless, the equipment developed for the project contributed directly to later successful tests on U-2 aircraft and thus to the eventual satellite program.

Optical degradation experiments determined that the shock wave, boundary-layer flow, and temperature gradients across windows in the bottom of the fuselage of X-15A-2 caused negligible degradation to visual, near-IR, and radar aerial photography to Mach 5.5 and 125,000 feet.

Ablator Tests: During the early 1960s, the only practical approach to speeds higher than Inconel X could withstand appeared to be an ablative coating of some description, much as was used on the early space capsules. Obviously, a better method of applying the ablator would have to be found, and it would need to be durable and maintainable. The material selected for use on the X-15 did not prove totally successful. Extensive man power was required to apply and refurbish the ablator surface, and then the integrity of the ablator-to-skin bonding was of concern for subsequent flights. Other operational problems argued against spray-on ablatives; the crew could not walk on the vehicle, and access panels were hard to remove and recover without leaving surface cracks. Also, many liquids, including liquid oxygen, would damage the ablator, requiring a coat of white paint to seal the ablative material's surface. The development of workable ceramic tiles (as used on the Space Shuttle) and metallic shingles (as proposed for some early Shuttle concepts; and now for X-33) have largely negated the need to use ablators. The short X-15A operational experience hastened the industry away from relying on ablators for reusable space vehicles.

Hypersonic Research Engine

With little doubt, the most ambitious[47] of the X-15 experiments was the Hypersonic Research Engine (HRE) from the Langley Research Center. At the time that researchers began to consider supersonic-combustion ramjet engines during 1954, the X-15 was not an approved program and played no major role in the engine's conceptual development. However, events soon transpired that made flight testing of a supersonic ramjet engine desirable, and the Flight Research Center and Langley proposed a joint project to accomplish just that. The 1962 crash of the X-15-2 opened the door for extensive modification aimed primarily at providing a platform for development of the Mach 8 air-breathing HRE. Then, as now, no tunnel facility existed wherein such an engine could be realistically tested, and rocket boosters could not give steady-state tests or return the equipment.[48]

The actual prototype engine was to be carried attached to the lower ventral of the X-15A-2. Twenty-nine inches were added to the fuselage between the existing tanks for the liquid hydrogen to power the HRE, two external fuel tanks were added, and the entire aircraft was coated with an ablative-type insulator.

During 1965, Garrett-AirResearch was put under contract to provide six prototype engines by mid-1969. As would happen, the development effort necessary to produce a workable engine had been severely underestimated, and Garrett quickly ran into problems that caused serious delays in the project.

In the meantime flight-test evaluations were made of the modified aircraft itself and of a dummy HRE attached to the X-15A-2. On the first and only maximum-speed test of the X-15A-2 in 1967, shock impingement off the dummy HRE caused severe heating damage to the lower empennage, and very nearly resulted in loss of the aircraft. Though quickly repaired, the X-15A-2 never flew again. Hindsight would place the blame for this design oversight on haste and insufficient flow interaction studies. A key lesson learned from this episode was not to hang external stores or pylons on hypersonic aircraft, at least not without far more extensive study of underside flow patterns. As John Becker later observed, "Flight testing on the X-15A-2 would have been long-delayed, hazardous, very costly, and fortunately never came about."[49]

When the X-15 flight program was terminated, the HRE degenerated into a costly wind tunnel program using partial-simulation test models. The HRE was eventually tunnel tested in 1969, and the primary objective of achieving supersonic combustion was met, although the thrust produced was less than the drag created. HRE engineers nonetheless claim a success in that the objective was supersonic combustion, not a workable engine. The program continued until 1975 and never achieved a positive net thrust, although it still contributed to the technology base, albeit at a very high cost. A hindsight study conducted in 1976 concluded that the HRE's fuel-cooled structure was its main contribution to future scramjets.[50]

Papers Published

Not the least of the technological legacies of the X-15 consisted of the more than 765 technical documents produced in association with the program, including some 200 papers reporting on general research that the X-15 inspired. John Becker saw them as "confirmation of the massive stimulus and the focus provided by the [X-15] program."[51]

Other Views

William Dana took time in 1987 to write a paper for the Society of Experimental Test Pilots pointing out some of the lessons learned from the X-15 program.[52] Dana should know—he was the last pilot to fly the X-15. Two he cited were particularly appropriate to the designers of the X-30 and X-33, although neither heeded the lessons. They are included here in their entirety:

> The first lesson from the X-15 is: Make it robust. As you have already seen, the

X-15 was able to survive some severe mistreatment during landings and still came back to fly another day. The X-15 that broke up after a spinning re-entry had self-recovered from the spin prior to break up, and might well have survived the entire episode had fixed, rather than self-adaptive, damper gains been used during re-entry. Another example exists of where the X-15 did survive a major stress in spite of operating with a major malfunction. This flight occurred in June 1967, when Pete Knight launched in X-15 No. 1 on a planned flight to 250,000 feet. At Mach 4 and at an altitude of 100,000 feet during the boost, the X-15 experienced a complete electrical failure that resulted in shutdown of both auxiliary power units and, therefore loss of both hydraulic systems. Pete was eventually able to restart one of the auxiliary power units, but not its generator. By skillful use of the one remaining hydraulic system and the ballistic controls, Pete was able to ride the X-15 to its peak altitude of 170 or 180,000 feet, reenter, make a 180 degree turn back to the dry lake at Tonopah, and dead-stick the X-15 onto the lakebed. All of these activities occurred without ever flowing another electron through the airplane from the time of the original failure.

> There will be a hue and cry from some that the aerospace plane [X-30—NASP] cannot afford the luxury of robustness; that the aerospace plane, in order to be able to get to orbit, will have to be highly weight-efficient and will have to forego the strength and redundancy margins which allowed the X-15 to survive during adversity. And my answer to these people is: build your first aerospace plane with X-15 margins, even at the expense of performance; these margins will serve well while you are learning how to make your propulsion system operate and learning how to survive in the heating thicket of hypersonic flight. Someday, with this

knowledge in hand, it will be time to build a no-margins aerospace plane, but for now I suggest that you seize all the margins that you can because you will need them, as did the X-15.

The other lesson from the X-15 is: conduct envelope expansion incrementally. The typical increment of speed increase for the original X-15 was about half a Mach number. With this increment it was easy to handle the heating damage that occurred in the original speed expansion phase. Again, I would expect to hear protest from the aerospace plane community, because when using one-half Mach number increments it is a long flight test program to Mach 25. Indeed, I cannot specify what size bite to take during the aerospace plane envelope expansion, but I can offer you the X-15A experience, in which two consecutive flights carrying the dummy ramjet were flown to Mach numbers of 4.94 and 6.70. The former flight exhibited no heat damage because of the wake of the dummy ramjet; the latter flight almost resulted in the loss of the aircraft due to heat damage.

Looking at the X-33 program in particular, another lesson jumps out. There will only be a single X-33. The building of three X-15s allowed the flight test program to proceed even after accidents. In fact, each of the X-15s was severely damaged at some time or another requiring it to be rebuilt. Plus, with multiple aircraft, it is possible to have one aircraft down for modification while the others continue to fly. And should one aircraft be lost, as sometimes happens in flight research, the program can continue. In today's environment it is highly unlikely that the X-33 program would continue if it exploded during an engine test like the X-15-3 did while ground testing the XLR99. Hopefully the X-33 will not experience such a failure, but is that not part of the reason we conduct flight research—to learn from the failures as well as the successes?

The New Millennium

As we enter the new millennium, it is interesting to note how the X-15 has shaped aeronautics and astronautics. Indeed, when the X-33 program began during 1996, it was surprising to find that many of the younger contractor engineers were totally unaware of the X-15, and that most thought the SR-71 was the fastest aircraft that had ever flown, discounting the Space Shuttle. Interestingly, the young engineers at Dryden remembered the program, and when it came to setting up the instrumentation range (which extends all the way to the Dakotas), lessons learned from the X-15 High Range were used.[53]

The most obvious difference today has absolutely nothing to do with the technology of hypersonic flight. It is the political climate that surrounds any large project. The NASA Administrator, Daniel Goldin, told an X-33 all-hands meeting that it was "okay to fail"—a reference that many times in order to succeed, you first have to experience problems that appear to be failures. But this is not the climate that actually exists. Any failure is often used as an excuse to cut back or cancel a project. In most cases the only way to totally avoid failure is to completely understand what you are doing; but if you completely understood something, there would be no point in building an X-plane!

The X-15 is usually regarded as the most successful flight research program ever undertaken. But the program had its share of failures. The XLR99 destroyed the X-15-3 before it had even flown; but the aircraft was rebuilt and the XLR99 became a very successful research engine. On several occasions the X-15s made hard landings, sometimes hard enough to significantly damage the aircraft; each time they were rebuilt and flew again. Mike Adams was killed in a tragic accident; but less than four months later William Dana flew the next research flight. Yes, the X-15 failed often; but its successes were vastly greater.

Perhaps we have not learned well enough.

[1] In the foreword to Milton O. Thompson, *At the Edge of Space: the X-15 Flight Program* (Washington, DC: Smithsonian Institution Press, 1992), p. xii.

[2] John V. Becker, "Principal Technology Contributions of X-15 Program" (NASA Langley Research Center, 8 October 1968), in the files of the NASA History Office.

[3] Ronald G. Boston, "Outline of the X-15's Contributions to Aerospace Technology," typescript available in the NASA Dryden History Office. For those interested in Boston's original paper, the easiest place to find a copy is in the *Hypersonic Revolution*, recently republished by the Air Force History and Museums program. It constitutes the last section in the X-15 chapter.

[4] J. D. Hunley, "The Significance of the X-15," 1999, an unpublished typescript available at the DFRC History Office.

[5] Boston listed 1,300 degrees Fahrenheit as the maximum temperature, but William Dana reported 1,350 degrees Fahrenheit in his SETP and AIAA papers. Boston also listed the max-q as 2,000 psf, but in reality it was 2,202 psf on flight 1-66-111.

[6] John V. Becker, "The X-15 Program in Retrospect" (paper presented at the 3rd Eugen Sänger Memorial Lecture, Bonn, Germany, 4-5 December 1968), pp. 1-2.

[7] J. D. Hunley, "The Significance of the X-15," 1999, an unpublished typescript available at the DFRC History Office.

[8] John V. Becker, "The X-15 Program in Retrospect" (paper presented at the 3rd Eugen Sänger Memorial Lecture, Bonn, Germany, 4-5 December 1968), pp. 1-2.

[9] Ronald G. Boston, "Outline of the X-15's Contributions to Aerospace Technology," typescript available in the NASA Dryden History Office, pp. 12-15.

[10] Kay, W. D., *The X-15 Hypersonic Flight Research Program: Politics and Permutations at NASA*, in From Engineering Science to Big Science: The NACA and NASA Collier Trophy Research Project Winners, edited by Pamela E. Mack, NASA SP-4219, 1998, p. 163.

[11] This concept was finally dropped as the Shuttle development program moved into Phase B. See Dennis R. Jenkins, *The History of Developing the National Space Transportation System: The Beginning through STS-75* (second edition; Cape Canaveral, Florida: Dennis R. Jenkins, 1997), p. 85.

[12] The still-born advanced solid rocket motors (ASRM) would have had propellant grained shaped to reduce their thrust, eliminating the need for the SSMEs to be throttled.

[13] Charles J. Donlan, "The Legacy of the X-15" (a paper in the *Proceedings of the X-15 30th Anniversary Celebration*, Dryden Flight Research Facility, Edwards, California, 8 June 1989, NASA CP-3105), p. 96; Toll, Thomas A. and Fischel, Jack, *The X-15 Project: Results and New Research*, in Volume 2, No. 3 of Astronautics and Aeronautics, p. 24.

[14] John V. Becker, "Principal Technology Contributions of X-15 Program" (NASA Langley Research Center, 8 October 1968), in the files of the NASA History Office.

[15] Letter from A. Scott Crossfield to Dennis R. Jenkins, 30 June 1999.

[16] In the same letter Crossfield points out that "Pressure suit history is very badly chronicled." I definitely found this to be true, and it is a part of the final X-15 history that needs a great deal of attention paid to it.

[17] J. D. Hunley, "The Significance of the X-15," 1999, an unpublished typescript available at the DFRC History Office.

[18] Milton O. Thompson, *At the Edge of Space: the X-15 Flight Program* (Washington, DC: Smithsonian Institution Press, 1992), pp. 68-71, 154.

[19] John V. Becker, "The X-15 Program in Retrospect" (paper presented at the 3rd Eugen Sänger Memorial Lecture, Bonn, Germany, 4-5 December 1968), pp. 7-8; John V. Becker, "Principal Technology Contributions of X-15 Program" (NASA Langley Research Center, 8 October 1968), in the files of the NASA History Office.

[20] John V. Becker, "The X-15 Program in Retrospect" (paper presented at the 3rd Eugen Sänger Memorial Lecture, Bonn, Germany, 4-5 December 1968), pp. 8-9; Braslow, Albert L., *Analysis of Boundary-Layer Transition on X-15-2 Research Airplane*, NASA TN D-3487, 1966.

[21] John V. Becker, "The X-15 Program in Retrospect" (paper presented at the 3rd Eugen Sänger Memorial Lecture, Bonn, Germany, 4-5 December 1968), pp. 9-10.

[22] J. D. Hunley, "The Significance of the X-15," 1999, an unpublished typescript available at the DFRC History Office.

[23] Ronald G. Boston, "Outline of the X-15's Contributions to Aerospace Technology," typescript available in the NASA Dryden History Office, p. 16-17; Milton O. Thompson, *At the Edge of Space: the X-15 Flight Program* (Washington, DC: Smithsonian Institution Press, 1992), pp. 70-71.

[24] Ronald G. Boston, "Outline of the X-15's Contributions to Aerospace Technology," typescript available in the NASA Dryden History Office, p. 18; J. D. Hunley, "The Significance of the X-15," 1999, an unpublished typescript available at the DFRC History Office.

[25] J. D. Hunley, "The Significance of the X-15," 1999, an unpublished typescript available at the DFRC History Office.

[26] On a corner of the impact range at Edwards.

[27] J. D. Hunley, "The Significance of the X-15," 1999, an unpublished typescript available at the DFRC History Office.

[28] Ibid.

[29] Ibid.

[30] There is often a great deal of confusion between the Mercury Control Center at Cape Canaveral used during the Mercury and initial Gemini flights, and the Mission Control Center in Houston that has been used since Gemini 5. Both, confusingly, are abbreviated MCC.

[31] J. D. Hunley, "The Significance of the X-15," 1999, an unpublished typescript available at the DFRC History Office.

[32] Ibid.

[33] Ronald G. Boston, "Outline of the X-15's Contributions to Aerospace Technology," typescript available in the NASA Dryden History Office, pp. 18-19; Milton O. Thompson, *At the Edge of Space: the X-15 Flight Program* (Washington, DC: Smithsonian Institution Press, 1992), pp. 182-186.

[34] The T-38 had modified speed brakes and a tweaked flight control system that allowed it to fly the steep approaches used by the orbiter.

[35] Dennis R. Jenkins, *Lockheed SR-71/YF-12 Blackbird*, (WarbirdTech Series Volume 10, Specialty Press, 1999), p. 37-38; Dennis R. Jenkins, *The History of Developing the National Space Transportation System: The Beginning through STS-75* (second edition; Cape Canaveral, Florida: Dennis R. Jenkins, 1997), p. 179.

[36] Ronald G. Boston, "Outline of the X-15's Contributions to Aerospace Technology," typescript available in the NASA Dryden History Office, pp. 16, 20.

[37] J. D. Hunley, "The Significance of the X-15," 1999, an unpublished typescript available at the DFRC History Office.

[38] Ibid., p. 11.

[39] Ronald G. Boston, "Outline of the X-15's Contributions to Aerospace Technology," typescript available in the NASA Dryden History Office, p. 17; Burke, Melvin E., *X-15 Analog and Digital Inertial Systems Flight Experience*, NASA Technical Note D-4642, 1968, pp. 1-2, 19.

[40] J. D. Hunley, "The Significance of the X-15," 1999, an unpublished typescript available at the DFRC History Office.

[41] Ibid.

[42] Wendell H. Stillwell, *X-15 Research Results*, (Scientific and Technical Information Branch, NASA, Washington, DC.: NASA SP-60, 1965), pp. 78-79; Milton O. Thompson, *At the Edge of Space: the X-15 Flight Program* (Washington, DC: Smithsonian Institution Press, 1992), p. 209.

[43] Milton O. Thompson, *At the Edge of Space: the X-15 Flight Program* (Washington, DC: Smithsonian Institution Press, 1992), pp. 87 and 188; Wendell H. Stillwell, *X-15 Research Results*, (Scientific and Technical Information Branch, NASA, Washington, DC.: NASA SP-60, 1965), p. 79.

[44] Milton O. Thompson, *At the Edge of Space: the X-15 Flight Program* (Washington, DC: Smithsonian Institution Press, 1992), pp. 188, 210, 263; John V. Becker, "The X-15 Program in Retrospect" (paper presented at the 3rd Eugen Sänger Memorial Lecture, Bonn, Germany, 4-5 December 1968), p. 6; J. D. Hunley, "The Significance of the X-15," 1999, an unpublished typescript available at the DFRC History Office.

[45] Ronald G. Boston, "Outline of the X-15's Contributions to Aerospace Technology," typescript available in the NASA Dryden History Office, p. 20.

[46] Dennis R. Jenkins, *Lockheed SR-71/YF-12 Blackbird*, (WarbirdTech Series Volume 10, Specialty Press, 1999), p. 37-38; Jenkins, Dennis R., *Space Shuttle: The History of Developing the National Space Transportation System*, 1996, p. 224.

[47] Letter from John V. Becker to Dennis R. Jenkins, 10 January 2000.

[48] John V. Becker, *A Hindsight Study of the NASA Hypersonic Research Engine Project*, unpublished study conducted under NASA Contract NAS1-14250, 1 July 1976. Typescript available in the files of the DFRC History Office.

[49] Letter from John V. Becker to Dennis R. Jenkins, 10 January 2000.

[50] John V. Becker, *A Hindsight Study of the NASA Hypersonic Research Engine Project*, unpublished study conducted under NASA Contract NAS1-14250, 1 July 1976. Typescript available in the files of the DFRC History Office.

[51] J. D. Hunley, "The Significance of the X-15," 1999, an unpublished typescript available at the DFRC History Office.

[52] Letter, William H. Dana, Chief, Flight Crew Branch, DFRC, to Lee Saegesser NASA History Office, transmitting a copy of the SETP paper for the file, in the files of the NASA History Office. A slightly rewritten (more politically correct) version of the paper was later published as *The X-15 Airplane—Lessons Learned* (American Institute of Aeronautics and Astronautics, a paper prepared for the 31st Aerospace Sciences Meeting, Reno Nevada, AIAA-93-0309, 11-14 January 1993)

[53] Personal observation of the author, who spent two years (off and on) helping the X-33 program.

Major Michael J. Adams poses in front of the X-15-1. Major Adams became the only fatality of the X-15 program when he was killed on Flight #191 while returning from high altitude. Adams was posthumously awarded Astronaut Wings for his last flight. (Tony Landis Collection)

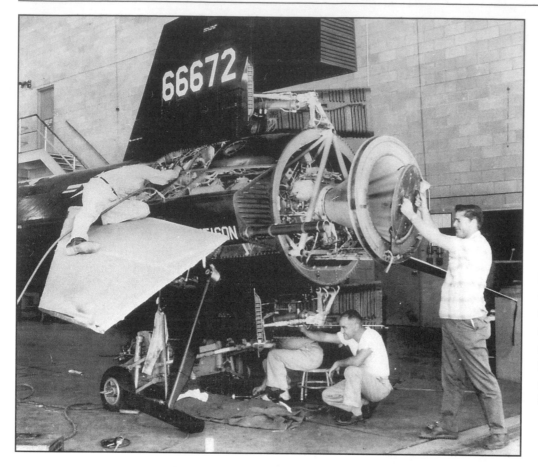

Technicians at the Flight Research Center work on the XLR99 engine. Note the corrugations on the aft end of the fuselage sponsons and vertical stabilizer. This was one of the keys to allowing the X-15 to withstand the high temperatures encountered during hypersonic flight. The blunt ends of the verticals and fuselage tunnels alone created as much drag as experienced by an F-104 fighter. (San Diego Aerospace Museum Collection)

A great deal of X-15 research did not involve the actual aircraft. Here a rocket sled is being used to test the ejection system. (Jay Miller Collection)

Appendix 1

Resolution Adopted by NACA Committee on Aerodynamics, 5 October 1954

This resolution was the official beginnings of the X-15 research airplane program.

**RESOLUTION ADOPTED BY NACA
COMMITTEE ON AERODYNAMICS, 5 OCTOBER 1954**

WHEREAS, The necessity of maintaining supremacy in the air continues to place great urgency on solving the problems of flight with man-carrying aircraft at greater speeds and extreme altitudes, and

WHEREAS, Propulsion systems are now capable of propelling such aircraft to speeds and altitudes that impose entirely new and unexplored aircraft design problems, and

WHEREAS, It now appears feasible to construct a research airplane capable of initial exploration of these problems,

BE IT HEREBY RESOLVED, That the Committee on Aerodynamics endorses the proposal of the immediate initiation of a project to design and construct a research airplane capable of achieving speeds of the order of Mach Number 7 and altitudes of several hundred thousand feet for the exploration of the problems of stability and control of manned aircraft and aerodynamic heating in the severe form associated with flight at extreme speeds and altitudes.

Appendix 2

Signing the Memorandum of Understanding

~~SECRET~~

DEPARTMENT OF THE AIR FORCE

WASHINGTON

OFFICE OF THE SECRETARY

NOV 9 1954

MEMORANDUM FOR THE ASSISTANT SECRETARY OF THE NAVY FOR AIR

SUBJECT: Principles for the Conduct of a Joint Project for a New
High Speed Research Airplane

    1. The Air Force concurs in the establishment of a joint NACA-
Navy-Air Force project to design and construct a research airplane
capable of achieving speeds of the order of Mach Number 7 and altitudes
of several hundred thousand feet.

    2. Attached is a Memorandum of Understanding, signed in tripli-
cate by the Air Force, setting forth the principles for the conduct
by the NACA, the Navy, and the Air Force of this joint project. It
is requested that the Navy sign this Memorandum, in triplicate, and
forward the signed copies to the Director of the NACA for signature
and distribution back to the signatory agencies.

    3. The Air Force is designating Brigadier General B. S. Kelsey,
Deputy Director of Research and Development, as the Air Force representative on the "Research Airplane Committee" referred to in paragraph B
of the Memorandum of Understanding.

(signed)
Trevor Gardner
Special Assistant (R&D)

Enclosure
   Memo of Understanding
   w/1 incl (in trip)

The first of three let-
ters attached to the
Memorandum of
Understanding that
created the X-15
research program.
Since it was nominally
an Air Force program,
the Air Force began
the signature process.

The early 1950s was
an era where carbon
paper and onion-skin
copies were kept.
Forty-five years later
they are not repro-
ducible, so the three
letters have been
recreated.

The letters remained
SECRET until 3 July
1963 when they were
downgraded to
CONFIDENTIAL.
It was not until 9
November 1966 that
they were finally
declassified.

The Navy was next to sign the Memorandum of Understanding. The letter is not dated, but other sources list it as being sent on 21 December 1954.

~~SECRET~~

DEPARTMENT OF THE NAVY
OFFICE OF THE SECRETARY
WASHINGTON

022421

Dear Doctor Dryden:

The enclosed copy of a Department of the Air Force memorandum of 9 November 1954 signed by Mr. Trevor Gardner, Special Assistant (R & D) expresses the Air Force concurrence in the establishment of a joint NACA-Navy-Air Force project to design and construct a research airplane capable of achieving speeds of the order of Mach Number 7 and altitudes of several hundred thousand feet. The Department of the Navy also concurs in the establishment of this joint project.

The enclosed Memorandum of Understanding, signed in triplicate by the Navy and the Air Force, sets forth the principles for the conduct by the NACA, the Navy and the Air Force of this joint project. This Memorandum of Understanding is forwarded for signature by the Director of the NACA and for distribution back to the signatory agencies.

RADM R. S. Hatcher USN, Assistant Chief for Research and Development, Bureau of Aeronautics, is designated as the Navy representative on the "Research Airplane Committee" referred to in paragraph B of the Memorandum of Understanding. The Air Force representative on this committee is designated in the enclosed Department of the Air Force Memorandum.

                                        Sincerely yours,

                                            (signed)
                                          J.H. Smith, Jr
                                Assistant Secretary of the Navy (Air)

Dr. Hugh L. Dryden
Director, national Advisory Committee for
Aeronautics
1512 H Street, N.W.
Washington 25, D.C.

Encl:
   Copy of Department of the Air Force
     Memorandum of 9 Nov 1954
   Memorandum of Understanding (in triplicate)

~~SECRET~~

NATIONAL ADVISORY COMMITTEE FOR AERONAUTICS
WASHINGTON

December 23, 1954

Mr. Trevor Gardner
Special Assistant for Research and Development
Department of the Air Force
4E964 National Defense Building
Washington 25, D.C.

Dear Mr. Gardner:

      The National Advisory Committee for Aeronautics con-
curs in the establishment of a joint NACA-Navy-Air Force project
to design and construct a research airplane capable of achieving
speeds of the order of mach Number 7 and altitudes of several
hundred thousand feet.

      The Memorandum of Understanding setting forth the
principles for the conduct of this joint project has now been signed,
in triplicate, by the Air Force, Navy, and NACA. A signed copy is
forwarded to you herewith.

      The "Research Airplane committee" referred to in
paragraph B of the Memorandum of Understanding is composed of
the following members:

        Brigadier General B.S. Kelsey, USAF, Deputy Director,
          Research and Development, U.S. Air Force
        Rear Admiral R.S. Hatcher, USN, Assistant Chief for
          Research and Development, Navy Bureau of Aeronautics
        Dr. Hugh L. Dryden, Director, National Advisory Commit-
          tee for Aeronautics

Sincerely yours,

(signed)
Hugh L. Dryden
Director

Enc.
HLDbk1

Hugh Dryden, from NACA, was the final signature, on the last working day of 1954.

This set in motion a chain of events that would lead to the design of the fastest manned aircraft yet conceived, and the construction of three flight research vehicles.

The first of these would fly less than five years later.

The Memorandum of Understanding that set up the "Research Airplane Committee" and established the workings of the X-15 research program.

The 5 October 1954 recommendation from the NACA Committee on Aeronautics was attached as a reference.

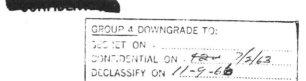

MEMORANDUM OF UNDERSTANDING

SUBJECT: Principles for the Conduct by the NACA, Navy and Air Force of a Joint Project for a New High Speed Research Airplane

A. A project for a high speed research airplane shall be conducted jointly by the NACA, the Navy and the Air Force to implement the recommendations of the NACA Committee on Aerodynamics, as adopted on 5 October 1954, copy attached.

B. Technical direction of the project will be the responsibility of the Director, NACA, acting with the advise and assistance of a "Research Airplane Committee" composed of one representative each from the NACA, Navy and Air Force.

C. Financing of the design and construction phases of the project shall be determined jointly by the Navy and Air Force.

D. Administration of the design and construction phases of the project shall be performed by the Air Force in accordance with the technical direction as prescribed in paragraph B.

E. The design and construction of the project shall be conducted through a negotiated contract (with supplemental prime or sub-contracts) obtained after evaluating competitive proposals invited from competent industry sources. The basis for soliciting proposals will be the characteristics determined by the configuration on which the NACA has already done much preliminary design work.

F. Upon acceptance of the airplane and its related equipment from the contractor, it will be turned over to the NACA, who shall conduct the flight tests and report the results of same.

G. The Director, NACA, acting with the advise and assistance of the Research Airplane Committee, will be responsible for making periodic progress reports, calling conferences, and disseminating technical information regarding the progress and results of the project by other appropriate media subject to the applicable laws and executive orders for the safeguarding of classified security information.

Memorandum of Understanding, "Principles for the Conduct by the **NACA**, Navy and **Air** Force of a Joint Project for a New High **Speed Research Airplane**"

H.   Accomplishment of this project is a matter of **national urgency**.

1 Incl
  Resolution Adopted by
NACA Committee on
Aerodynamics, 5 Oct 54

                                        Hugh L. Dryden   NACA
                                            Director, **NACA**

                                     J. H. Smith, Jr.   Navy
                           Assistant Secretary of the Navy (**Air**)

                                     Trevor Gardner   AirForce

2

RESOLUTION ADOPTED BY NACA
COMMITTEE ON AERODYNAMICS, 5 OCTOBER 1954

WHEREAS, The necessity of maintaining supremacy in the air continues to place great urgency on solving the problems of flight with man-carrying aircraft at greater speeds and extreme altitudes, and

WHEREAS, Propulsion systems are now capable of propelling such aircraft to speeds and altitudes that impose entirely new and unexplored aircraft design problems, and

WHEREAS, It now appears feasible to construct a research airplane capable of initial exploration of these problems,

BE IT HEREBY RESOLVED, That the Committee on Aerodynamics endorses the proposal of the immediate initiation of a project to design and construct a research airplane capable of achieving speeds of the order of Mach Number 7 and altitudes of several hundred thousand feet for the exploration of the problems of stability and control of manned aircraft and aerodynamic heating in the severe form associated with flight at extreme speeds and altitudes.

Appendix 3

Preliminary Outline Specification

The preliminary specification for the future X-15 did not differ substantially from the final version published a few weeks later. Although engines were not specifically discussed in the written document, the aircraft depicted in Figure 2 was powered by three Hermes A3A engines and assumed a launch by a modified B-50 carrier aircraft.

~~CONFIDENTIAL~~

PRELIMINARY OUTLINE SPECIFICATION

FOR

HIGH-ALTITUDE, HIGH-SPEED RESEARCH AIRPLANE

October 15, 1954

1.   STATEMENT OF PROBLEM AND OBJECTIVE:

la   The next major advance in aircraft performance will plunge the aircraft designer into a speed range where the accompanying temperature effects would cripple the strength of conventional aircraft materials and structures and into an altitude range where the air pressure is too low for conventional aerodynamic controls. In addition, certain physiological and environmental problems associated with the pilot of such a high-speed high-altitude airplane are anticipated. Many of the most important problems in this field can be satisfactorily investigated only with a manned full-scale flight vehicle.

lb   In order to provide the fundamental research information essential to the practical solution of these problems in this country, a need exists for a research airplane capable of exploring the speed and altitude regimes in which these problems are encountered.

lc   As the need for the exploratory data is acute because of the rapid advance of the performance of service aircraft, the minimum practical and reliable airplane is required in order that the development and construction time be kept to a minimum.

2.   APPLICABLE SPECIFICATIONS, STANDARDS, DRAWINGS, AND OTHER PUBLICATIONS:

2.la General specifications for the design and construction of airplanes for the United States Navy, SD-24G dated   shall be followed where applicable.

2.lb Materials, process, design and installation specifications and equipment drawings, applicable to piloted aircraft in effect as of this date shall be followed where applicable.

2.lc Deviations from applicable Government specifications and standards will be encouraged provided these deviations are directed toward the accomplishment of the objective set forth in paragraph lc.

2.2a Specifications and standards shall be used in the order of precedence set forth in ANA Bulletin 143c.

~~CONFIDENTIAL~~

~~CONFIDENTIAL~~

- 2 -

3.   REQUIREMENTS:

3.1a   GENERAL - Major considerations that established the required
       performance of the airplane are:

(a) For exploring the aerodynamic heating problem, the structure
must be subjected to extreme heating conditions. Allowable
skin temperatures are of the order of 1200° F and maximum
heating rates of the order of thirty (30) BTU/sq. ft/sec.
are desired. Altitude-speed requirements are also such that
radiation heat loss is of comparable magnitude to the con-
vective heat input with resultant skin temperatures well
below adiabatic boundary-layer temperature.

(b) For exploring the stability and control problems of a manned
high-altitude aircraft, the speed-altitude capabilities of the
research airplane should permit the establishment of flight
conditions for which aerodynamic forces are negligible compared with inertia forces, thus requiring the use of auxiliary
controls.

(c) For exploring physiological factors affecting pilot response,
the research airplane should be capable of effecting periods of
"weightlessness" for a long enough period to permit exploration
of this field.

(d) Provisions should be made to substitute an observer in the
space alloted for research instrumentation.

3.2a   PERFORMANCE

(a) The airplane shall be capable of achieving a speed of at
least 6600 ft/sec.

(b) The airplane shall be capable of attaining an altitude of at
least 250,000 feet.

3.2b   AIRPLANE WEIGHT AND SIZE - The size and weight of the airplane shall
       be such as to permit air launching from a mother airplane: such as
       the B-50, B-36, or B-52, thus effectively providing a two-stage
       vehicle.

3.3a   OPERATIONAL FACTORS - The research airplane will normally be operated
       from and in the vicinity of the Edwards Air Force Base, California.
       The presence of the large landing areas afforded by the dry lakes in
       the vicinity may be taken into consideration in the design of the
       airplane for the landing phase of the flights.

~~CONFIDENTIAL~~

~~CONFIDENTIAL~~

- 3 -

3.4a VISION - A reasonable degree of vision, direct or not, should be afforded the pilot particularly in the landing approach attitude.

3.4b Vision for the observer shall be provided only to the extent that scientific observation may be satisfied.

4. STABILITY AND CONTROL:

4.1a The flying qualities and general handling characteristics of the airplane during all phases of the flight, both inside and outside of the atmosphere shall be adequate to permit satisfactory ful-fillment of the mission and utilization as specified herein.

4.1b The wing and tail arrangement shall be such as to offer promise in the light of existing aerodynamic knowledge, of attaining good stability and control characteristics throughout angle-of-attack range at low speeds, as well as at high speeds.

4.1c Controls shall be provided to permit changing airplane attitude in the absence of aerodynamic forces.

4.1d Where an artificial feel system is employed, the system shall be foolproof, reliable, and as simple as possible consistent with the force requirements. Any complicated and/or apparently unreliable system shall be unacceptable.

4.1e Through combination of aerodynamic features, such as powerful dive brakes and/or large drag at high angles of attack, and structural features, such as thick skin, auxiliary cooling, and/or high temperature alloys, it shall be possible to recover from flights to maximum speed or maximum altitude without exceeding the allowable temperature limits for the structure or the accel-erations currently encountered in combat with fighters.

5. STRUCTURAL DESIGN CRITERIA:

5.1a The high temperatures and aerodynamic heating loads anticipated in the operating regime of the airplane require that careful attention be given to the choice of structural materials and/or to methods for cooling and/or insulating the surfaces.

5.1b The design normal loads shall be +7.50g and -3g or the A.F. fighter equivalent.

~~CONFIDENTIAL~~

~~CONFIDENTIAL~~

- 4 -

6.   FURNISHINGS AND EQUIPMENT:

6.1a Provisions for cockpit pressurization and air conditioning shall
      be adequate for flights to maximum speed and altitude.

6.1b Provisions shall be made to permit the use by the pilot of full
      pressure suit. The pilot shall have reasonable protection from
      radiated and conducted heat.

6.1c Suitable escape provisions shall be provided for the pilot.

6.1d Provisions for breathing oxygen shall be sufficient for the complete
      flight.

6.2a The observer shall be provided with protection and escape provisions
      equal to those provided for the pilot.

6.3a All instruments necessary for the proper performance of the airplane
      shall be provided for the pilot.

7.   PROPULSION:

7.1a The propulsion system chosen shall be suitable for a manned aircraft.
      A list of powerplants which with reasonable development may be used
      for the project follows:

          (List to be provided by BuAer and WADC)

8.   RESEARCH INSTRUMENTATION:

8.1a A weight allowance of 500 pounds and a volume allowance of 5 cubic
      feet shall be provided for research instrumentation. Provision
      for pressurization and cooling must also be made. Thermocouples
      shall be installed for determining the temperature distribution
      throughout the airframe.

~~CONFIDENTIAL~~

- 5 -

APPENDIX

USE OF NACA FACILITIES FOR FINAL DEVELOPMENT:

High Mach number wind tunnel and structural development work are essential to establish the final design of such a research airplane. Facilities for such work are in existence at NACA Laboratories and will be made available for development of the selected design.

SUGGESTED MEANS OF MEETING GENERAL REQUIREMENTS:

The NACA has made studies to determine if, on basis of the existing knowledge, it would be possible to develop and construct an airplane capable of meeting the preceding requirements. A typical flight plan is on figure 1. The airplane configuration evolved is shown in figure 2.

Figure 1 illustrates one of the flight trajectories that is possible with the airplane within the temperature limits of the structure. The airplane is launched from the mother ship at 35,000 feet. Burnout occurs at an altitude of 146,000 feet and at a speed of 6600 feet per second. In its subsequent ballistic trajectory, an altitude of 280,000 feet is achieved and for about 130 seconds in this trajectory the dynamic pressure is less than 6 pounds per square foot. During this period of time, the pilot will be required to change the attitude of the airplane from nose-up to nose-down as required for reentry using nonaerodynamic controls. In the reentry portion of the trajectory, the combined use of dive brakes and moderate lift on the airplane may be used to avoid excessive skin temperatures.

Figures 3 and 4 show schematically an internal wing structure which would permit thermal expansion of the wing without the production of large thermal stresses. Some of the more important features are noted in figure 2 or given below:

(a) Size and weight are such as to permit use of a B-50 mother ship for launching.

(b) Wing and tail arrangement offering promise of attaining good stability and control characteristics throughout angle-of-attack range at low speeds as well as at high speeds

(c) Split tail surfaces affording powerful means for providing required stability at very high speeds and avoiding the necessity for excessively large stabilizing surfaces

~~CONFIDENTIAL~~

- 6 -

(d) Split flaps on wing and tail surfaces to provide powerful dive brakes. Fuselage dive brakes may also be necessary.

(e) Rounded leading edge and leading-edge sweep of wing and tail surfaces, greatly reducing rate of heat transfer into these surfaces

(f) Skin thickness of 0.1-inch Inconel alloy, providing adequate heat sink to accomplish desired flight trajectories without exceeding a 1200° F temperature limit

(g) An interior web and rib detail minimizing the thermal stress problem by permitting free expansion of wing elements

(h) Use of skid type landing gear to avoid tire cooling problems

~~CONFIDENTIAL~~

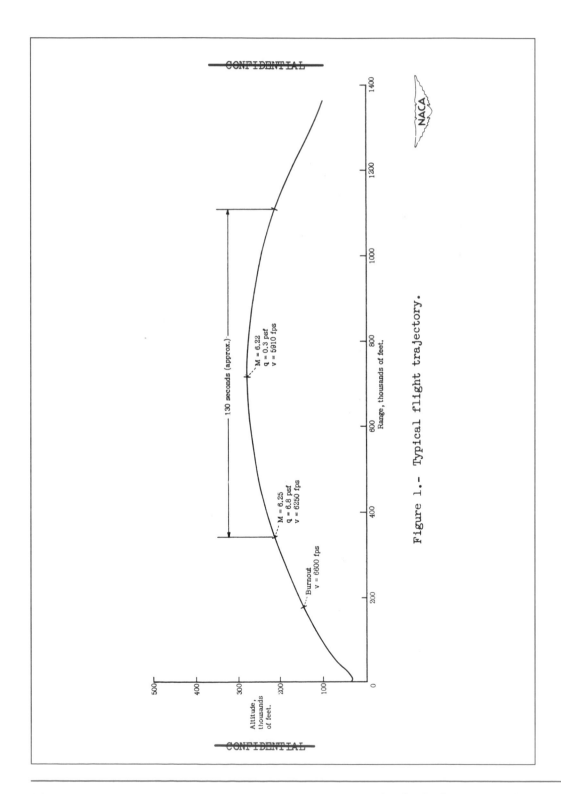

Figure 1.- Typical flight trajectory.

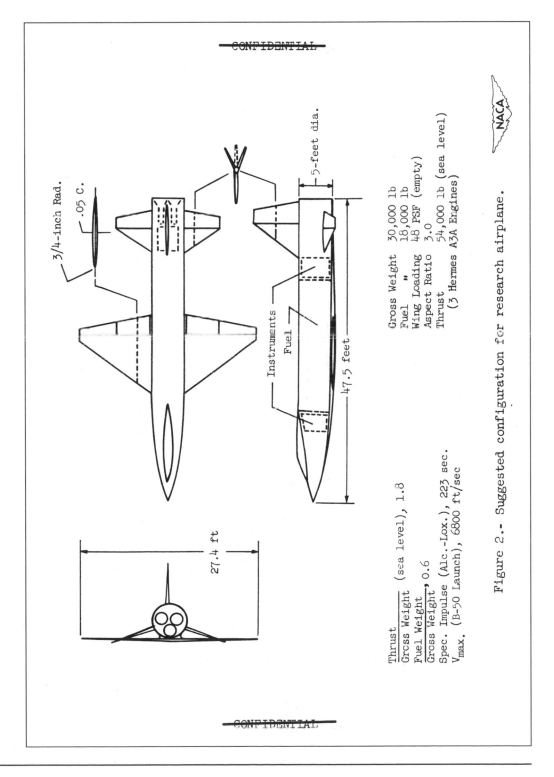

Figure 2.- Suggested configuration for research airplane.

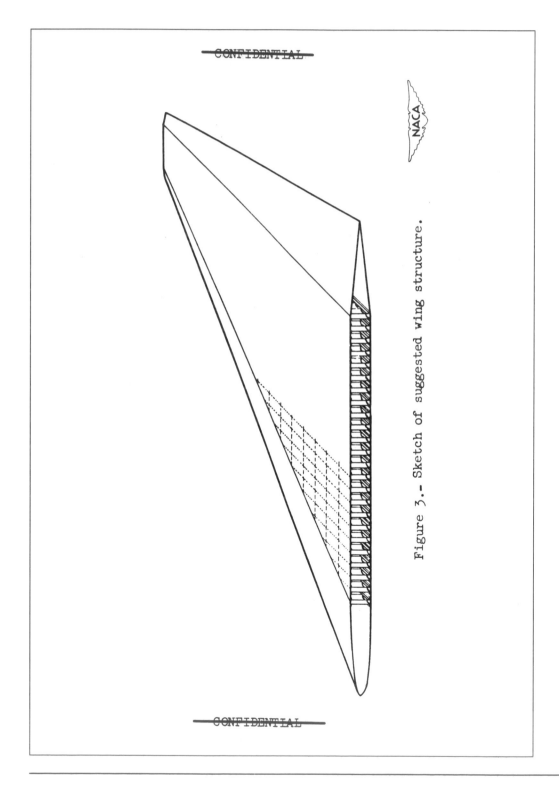

Figure 3.- Sketch of suggested wing structure.

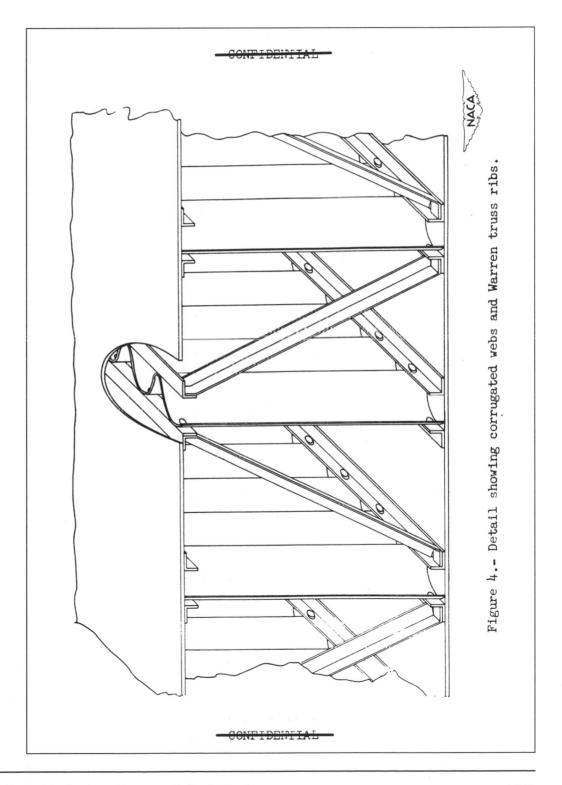

Figure 4.– Detail showing corrugated webs and Warren truss ribs.

Appendix 4

Surveying the Dry Lakes

NORTH AMERICAN AVIATION, INC.
INTER-OFFICE LETTERS ONLY

TO	G.R. Mellinger	DEPARTMENT	56
FROM	G. P. Lodge	DEPARTMENT	56
PHONE		DATE	1 December 1959
SUBJECT	Survey of Dry lakes in California, Nevada and Utah		

A survey was made of approximately 50 dry lakes in California, Nevada and Utah area to ascertain which lakes would be suitable for emergency landing sites for the X-15 airplane.

The method used for determining surface hardness was dropping an 18 pound, 5 inch diameter steel ball from a 6 foot height and measuring the diameter of the imprint in the surface. A diameter of less than 3 1/4 inches is considered satisfactory. In addition to the steel ball check, a 3/8 inch and a 1/2 inch diameter blunt end steel rod was also used to probe the surface to determine the thickness of the crust and soil condition under the crust which would have a direct effect upon load bearing qualities of the lake. A force of 200 pounds was applied to the rods and the depth of penetration measured.

Listed below are the lakes investigated and comments regarding the condition and use:

Location – 35° 17 N, 117° 28 W

Cuddeback Lake: Surface crust is moderately rough and damp in spots. Steel ball imprint varies from 3 to 4". 3/8 rod penetration up to 12". 1/2 rod 2 to 10".

This lake is considered marginal for emergency recovery.

Location – 35° 44 N, 117° 30 W

Searles lake: No landing made. Surface appeared soft and wet. Water and diggings on lake bed.

Not considered usable.

Location – 36° 00 N, 117° 14 W

Ballarat Lake: No landing made. Surface appeared soft and sandy with wet spots. One small area of the lake bed had checked surface. Road crosses north end.

Not considered usable.

Location – 36° 20 N, 117° 25 W

Panamint Springs: No landing made. Surface at south end was drifted sand. North end appeared hard but too small for X-15 use. A few ditches are on lake bed. Paved road crosses the north end.

Not considered usable for X-15.

A major task that needed completed before the first X-15 flight was a survey of available emergency and contingency landing areas along the projected flight corridor. Since the X-15 was equipped with skid-type landing gear, the only acceptable landing areas were dry lakebeds.

North American and the Air Force made several trips to survey the dry lakes along the flight corridor and to make tests on the most promising. The lakebed had to be smooth, long enough, and hard enough to accommodate the X-15.

G. R. Mellinger from G. P. Lodge   Page two
                                                    1 December 1959
- -

Location - 36° 30 N, 116° 55 W

Death Valley: No landings made. No usable spots noted.

Location - 36° 30 N, 116° 32 W

Scranton: No landings made. Water on surface.

Not considered usable.

Location - 36° 15 N, 116° 23 W

Death Valley Junction: No landing made. Surface appeared soft and numerous diggings
on lake bed.

Not considered usable.

Location - 36° 13 N, 116° 10 W

Stewart Valley: No landing made. Surface appeared soft and had drifted sand.

Not considered usable.

Location - 36° 17 N, 116° 03 W

Pahrump: No landing made. Surface appeared soft with drifted sand and brush
growing.

Not considered usable.

Location - 36° 00 N, 115° 57 W

Hidden Hills: Elevation 2,000 feet. Surface hard and smooth. Steel ball imprint
3 to 3 1/2 inches. 3/8 Rod penetration 3 1/4 to 3 1/2 inches. Water draining in
at north end. Approximately 15,000 feet of usable lake on 150° - 330° headings.
Approximately 2,000 to 3,000 feet more length available when dry. Access from
paved road 10 miles east.

This lake considered good for emergency recovery.

Location - 35° 43 N, 115° 35 W

Mesquite Lake: No landing made. Surface appeared soft with water and brush on
lake bed.

Not considered usable.

Location - 35° 32 N, 115° 22 W

Ivanpah Lake: Elevation 3,000 feet. Large lake with smooth moderately hard surface.
Steel ball imprint 3 1.4 to 3 1/2 inches. 3/8 Rod penetration 10 to 12 inches, 1/2
Rod 1 to 7 inches. Soil under crust was damp. Paved road crosses north end of lake
bed. Approximately 23,500 feet usable length on 160° - 340° headings.

This lake is considered marginal for emergency recovery.

```
G. R. Mellinger from G. P. Lodge                          Page three
                                                         1 December 1959
- - - - - - - - - - - - - - - - - - - - - - - - - - - - - - - - - - - -

Location - 35° 40  N, 115° 22  W

Reach Lake:  Smooth, slightly cracked surface.  Steel ball imprint 3 1/2 to 4 inches.
3/8 Rod penetration 15 inches.  Crust breaks up easily.  Soil loose under 1 inch
crust.  Railroad track crosses lake.

Not considered favorably for emergency recovery.

Location - 36° 27  N, 114.° 52  W

Dry Lake:  No landing made.  Surface too small for X-15 use.

Not considered usable for X-15.

Location - 36° 58  N, 115° 15  W

Cabin Springs:  No landing made.  Surface rough and uneven.  Brush on lake bed
around edges.

Not considered usable.

Location - 37° 20  N, 114° 55  W

Delamar Vallay:  Elevation 4,000 feet.  Surface moderately hard and smooth.  Dry
and hard under surface.  Steel ball imprint 3 inches.  3/8 Rod penetration 2 to
2 1/2 inches.  Usable length 13,500 feet on 0° - 180° headings.  Power line on S.E.
corner.  Wind was 10 - 15 mph from North.  Access to lake is from Alamo on U.S. 93.

This lake is considered good for emergency recovery.

Location - 37° 45  N, 114° 49  W

Dry Lake Valley:  No landing made.  Surface appeared soft with drifted sand.

Not considered usable.

Location - 37° 55  N, 115° 20  W

Coal Valley:  No landing made. Surface appeared to be soft sand.

Not considered usable.

Location - 38° 31  N, 115° 37  W

Currant Lake:  Surface rough and soft. Shallow ditches across center of lake bed.

Not considered usable.

Location - 39° 17  N, 115° 15  W

Jakes Lake:  Large lake.  Surface soft and rough with shallow ditches.  Grass growing
on lake bed and cattle grazing.

Not considered usable.
```

G. R. Mellinger from G. P. Lodge   Page four
                                                          1 December 1959
- -

Location - 39° 40 N, 115° 43 W

Newark Valley: Long lake. Surface smooth and soft. Steel ball imprint 4 1/2 inches.
Crust breaks up easily. Soil under crust damp.

Not considered usable.

Location - 40° 00 N, 115° 58 W

Diamond Lake: No landing made. Surface smooth and soft. Lake bed approximately
size of Rogers.

Not considered usable.

Location - 40° 23 N, 115° 25 W

Franklin Lake: Large lake partially covered with grass. Surface soft with loose
soil under soft crust. Steel ball imprint 4 1/2 inches. 3/8 Rod penetration 12
inches.

Not considered usable.

Location - 40° 08 N, 114° 42 W

Goshute Lake: No landing made. Surface soft, settle on lake bed.

Not considered usable.

Location - 41° 05 N, 113° 55 W

Area immediately east of Pilot Peak, surface white, smooth and soft. Steel ball
imprint 4 1/2 inches. 1/2 Rod penetration 12 inches plus. Soil under white salt
film wet.

Not considered usable.

Location - 40 ° 46 N, 113° 50 W

Bonneville Flat Race Track: Large area with white surface. Long black line on
headings of 30° - 210° marks course. Surface adjacent to line (1/4 to 1/2 mile
each side) exceptionally hard and composed of salt. Steel ball imprint 1 3/4 inches.
3/8 Rod penetration zero. Darker colored areas to each side soft. The area adjacent
to and parallel with the race track is considered an excellent emergency recovery
site.

Location - 40° 45 N, 114° 42 W

No landing made. Usable surface too small.

Not considered usable.

Location - 40° 47 N, 114 ° 57 W

Snow Water Lake: No landings made. Surface soft with water on west portion.

Not considered usable.

G. R. Mellinger from G. P. Lodge   Page five
                                                          1 December 1959
- -

Location - 40° 00 N, 116° 40 W

(Walti Hot Spring): Large lake; surface smooth, dry and soft. Steel ball imprint
4 to 4 1/2 inches. 3/8 Rod penetrated full length. Crust thin and breaks up easily.
Soil under crust dry and powdery.

Not considered usable.

Location - 40° 11 N, 116° 50 W

No name: No landing made. Surface appeared soft.

Not considered usable.

Location - 40° 25 N, 117° 20 W

No name: No landing made. Surface appeared soft. Cattle on lake bed had left deep
tracks. Top of cinder cone crater above surface at south end.

Not considered usable.

Location - 40° 14 N, 117° 58 W

Buena Vista Valley: Large lake. No landing made. Surface appeared soft, drifted
sand and deep cattle and car tracks on all portions of lake bed.

Not considered usable.

Location - 39° 20 N, 118° 30 W

Carson Sink: Vary large area. North portion covered with wide shallow ditches and
drifted sand. Landing area approximately 2 miles S.W. of target cone in N.E. portion
of lake. Surface smooth, checked and soft. Steel ball imprint 4 to 4 1/2 inches.
3/8 Rod penetration full length. Soil under crust damp and loose. Would not pack.
Touch and go landings made on other portions of lake bed indicated soft surface on
entire lake.

Not considered usable.

Location - 39° 20 N, 119° 25 W

No name: No landing made. Small lake with sandy brush covered surface.

Not considered usable.

Location - 39° 18 N, 119° 04 W

No name: No landing made. Brush covered surface.

Not considered usable.

Location - 39° 23 N, 118° 53 W

No name: Two lakes. No landing made. Surface appeared soft and brush covered.

Not considered usable.

G. R. Mellinger from G. P. Lodge   Page six
                                                        1 December 1959
- -

Location - 39° 09 N, 118° 42 W

No name: No landing made. Surface appeared soft and covered with ditches.

Not considered usable.

Location - 39° 22 N, 118° 36 W

No name: Small lake. No landing made. Surface appeared soft.

Not considered usable.

Location - 39° 19 N, 118° 30 W

No name: Large lake. No landing made. Surface appeared soft. Ditches on north
end of lake bed.

Not considered usable.

Location - 39° 16 N, 118° 16 W

Labou Flat: Small lake. No landing made. Road crosses surface and gunnery targets
installed on east side.

Not considered usable.

Location - 39° 37 N, 117° 39 W

No name: large lake. Elevation 5,000 feet. Mountains on east, north and west
sides. Wide valley to south. Lake bed is approximately 7 miles long on 30° - 210°
headings and 3 to 4 miles wide. Surface is smooth and moderately hard. Steel ball
imprint varied from 3 to 3 1/2 inches. 3/8 Rod penetration varied from 10 to 12
inches at north end and center to 1 to 3 inches in light colored area at south end.
Dark colored area at south end is soft. Boil 5 to 6 inches below crust dump. Best
touch-down point would be at south end in light colored area on 30° heading. Access
via dirt road from Eastgate, Nevada.

This lake is considered favorably for emergency recovery.

Location 39° 20 N, 117 ° 29 W

Smiths Ranch: Large lake with smooth hard surface. Elevation 5700 feet. Lake
bed is 7 to 8 miles long. Surface at south end is rougher but harder than north
end. Roughness is result of wider cracks that existed at one time in surface.
Steel ball imprint at north end 3 to 3 1/4 inches, center 2 3/4 to 3 inches, south
end 2 1/4 to 2 1/2 inches. 3/8 Rod penetration varied from 1 inch at north end,
2 1/2 or 3 1/2 inches at center to 1/2 or 3/4 inches at south end. Crust thickness
varied from 4 to 7 inches. Access is from paved road, U.S. 50 that runs adjacent to
lake bed.

This lake is considered an excellent emergency recovery site.

```
G. R. Mellinger from G. P. Lodge                    Page seven
                                                    1 December 1959
- - - - - - - - - - - - - - - - - - - - - - - - - - - - - - - - - - - - - - -
```

Location – 38° 54 N, 118° 15 W

Gabbs Valley: No landing made. Surface appeared soft, dark and wet.

Not considered usable.

Location – 38° 05 N, 117° 58 W

Columbus Salt Marsh: No landing made. Surface appeared soft and rough with ditches.

Not considered usable.

Location – 38° 01 N, 117° 38 W

Big Smoky Valley: Long narrow lake. Surface rough and uneven. Soft spots on
south half. Steel ball imprint at north end 3 1/4 to 3 1/2 inches. 3/8 Rod penetration full length. Crust thin and crumbles easily. Soil under crust loose. Approximately 10,000 to 15,000 feet of surface on headings of 20° - 200° at north end of
lake may be considered satisfactory for jet A/C.

Not considered usable for X-15.

Location 37° 52 N, 117° 23 W

Alkali Springs: Circular shaped lake. Surface smooth, cracked and hard. Elevation
4500 feet. Steel ball imprint 2 1/2 to 2 3.4 inches. 3/8 Rod penetration 1/2 inch.
Crust 4 inches thick. Soil under crust loose. Usable length 9,000 feet on 50° - 230°
heading.

This lake is considered excellent for emergency recovery of jet A/C but too small
for X-15.

Location – 37° 52 N, 117° 04 W

Mud Lake: Circular shaped lake. Elevation 5,000 feet. Surface smooth and hard.
Marked runways exist on headings of 60° - 240° and 170° - 350°. Steel ball imprint
varied from 2 to 3 inches. 3/8 Rod penetration was 1/8 to 1/4 inches except on
east side of lake where it could be pushed in all the way. The east portion of lake
is the softest part. Usable length of surface is 4 to 5 miles in any direction. It is
recommended that touch down not be made on east portion if possible. Access is by
dirt road approximately 10 miles from paved road.

This lake is considered usable for X-15.

Location – 37° 40 N, 117° 41 W

Clayton Valley: No landing made. Area covered with sand dunes.

Not usable.

Location – 37° 26 N, 117° 09 W

No name: No landing made. Surface covered with sand and brush except for small open
area.

Not considered usable

G. R. Mellinger from G. P. Lodge

Page eight
1 December 1959

- -

Location - 37° 13 N, 117° 05; W

No name: No landing made. Surface covered with volcanic mud flow. Major portion of surface wet.

Not considered usable.

Location - 37° 10 N, 117° 10 W

No name: Small lake. Surface smooth and hard. Steel ball imprint 2 1/2 to 2 3/4 inches. 3/8 Rod penetration 2 to 3 inches. Usable length 10,000 feet on 10° - 190° heading. Numerous small brush covered islands scattered on surface.

This lake is considered usable for emergency recovery of jet A/C but too small for X-15.

The attached sketch shows the location of the lakes inspected by latitude and longitude. Also included on the sketch are lake beds previously inspected by the writer and those inspected by L/Col. Anderson and Major White of Edwards Flight Test Center. The lake beds designated "most usable" were selected from the standpoint of size, surface conditions and access for recovery of vehicle.

                                                    G. P. Lodge
                                                    Flight Safety Specialist

GPL:lr

Dc: Ferren, Crossfield, White, Roberts, Wilkerson, Helgeson, Cokeley, Harvey,
     Richter, Stacey, Beach, Jelinek, O Conner, Lodge, File (10).

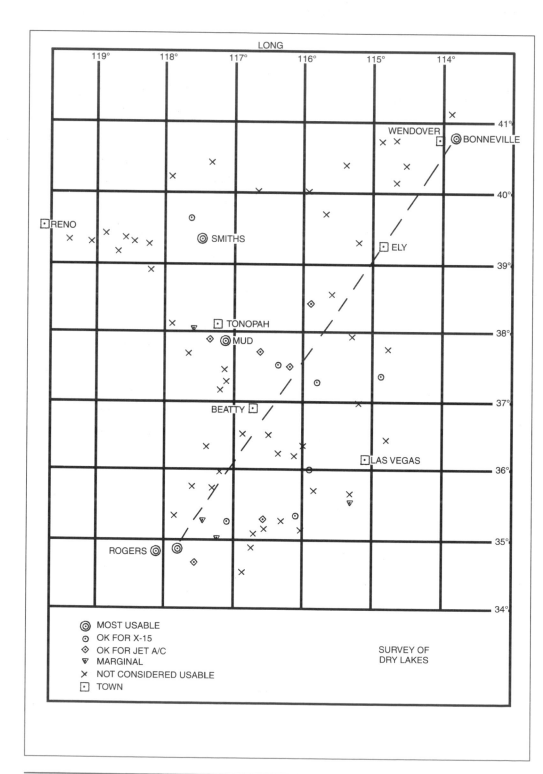

This recreation of the original sketch shows the location of the lakes inspected by latitude and longitude. Also included on the sketch are lake beds previously inspected by Mr. Lodge and those inspected by Lieutenant Colonel Anderson and Major White of the Air Force Flight Test Center at Edwards AFB.

The lake beds designated "most usable" were selected from the standpoint of size, surface conditions and access for recovery of vehicle.

R&D Project Card—Project 1226, X-15 Research Aircraft

The Air Force's Air Research and Development Command (ARDC) was the lead organization for the development and procurement of the X-15 airplanes. This "Project Card" initiated the paperwork for the project. Like much of the early X-15 data, it was classified SECRET.

SECRET
SECURITY CLASSIFICATION

R&D PROJECT CARD	TYPE OF REPORT New Project	REPORT CONTROL SYMBOL DD-RDB(A)48

1. PROJECT TITLE (CONFIDENTIAL) X-15 Research Aircraft	2. SECURITY OF PROJECT SECRET	3. PROJECT NUMBER 1226
	4. 1-1226	5. REPORT DATE 7 March 1955

6. BASIC FIELD OR SUBJECT Technical Development	7. SUBFIELD OR SUBJECT SUBGROUP 01-Aircraft and Design Studies	7A. TECH OBJ SR-1f

8. COGNIZANT AGENCY ARDC	12. CONTRACTOR AND/OR LABORATORY Contractor to be selected after a Design Competition	12A. CONTRACT-W.O. NO.
9. DIRECTING AGENCY Fighter Aircraft Division, WADC	This material contains information affecting the national defense of the United States within the meaning of the Espionage Laws, Title 18, U.S.C. Sections 793 and 794, the transmission or revelation of which in any manner to an unauthorized person is prohibited.	

9A. OFFICE SYMBOL WCSFF	9B. EXTENSION 39159

10. REQUESTING AGENCY NACA - Hq USAF	13. RELATED PROJECTS X-1A, X-1B, X-1E, X-2	17. EST COMPL DATES RES. Continuing DEV. Dec 1958 TEST Mar 59(Phase II) OP. EVAL.
11. PARTICIPATION, COORDINATION, INTEREST US Navy (P) NACA (P)	14. DATE APPROVED 7 March 1955	18. FY See Item 21d(2) FISCAL ESTS (M$)
	15. PRIORITY 1-B	16. Category A-1

19.

20 REQUIREMENT AND/OR JUSTIFICATION
As a result of studies made by the NACA between June 1952 and July 1954, it was concluded that the two most serious problems which will be encountered in flight at very high speeds and altitudes are: (1) prevention of the destruction of the aircraft structure by the direct or indirect effects of aerodynamic heating and, (2) achievement of satisfactory stability and control. A review of existing and planned facilities suitable for these investigations indicates that while certain phases of these problems can be studied in the laboratory, there will remain many questions which can only be answered by full scale flight research. A manned airplane was considered to be feasible since the nature of the stability and control problem will dictate that the high performance be attained by moderate incremental increase starting from speeds and altitudes at which information is already available. Technical Program Requirement No. 1-1 dated 6 October 1954 was established by Hq USAF with the

01

22. R&D	SN	CN	IC&P	X.	I.	C.

DD FORM 613
1 JAN 52

PREVIOUS EDITIONS OF THIS FORM MAY BE USED

SECRET
SECURITY CLASSIFICATION
SECRET

PAGE 1 OF 4 PAGES
C5-38426

R&D PROJECT CARD
CONTINUATION SHEET
SECRET
SECURITY CLASSIFICATION

SECRET

1. PROJECT TITLE	2. SECURITY OF PROJECT	3. PROJECT NUMBER
(CONFIDENTIAL) X-15 Research Aircraft	SECRET	1226
	4. 1-1226	5. REPORT DATE 7 March 1955

20. (contd)

objective of initiating a new manned research airplane project generally **in**
accordance with the NACA Secret report, subject, "NACA Views Concerning a New
Research Aircraft," dated August 1954. (Conf)

21.a. Brief

This project is being initiated to develop an air launched, rocket
propelled, manned aircraft capable of flight at speeds of at least 6600 ft/sec.
and altitudes of at least 250,000 ft. It has been generally accepted that the
state-of-the-art will support the development of a vehicle capable of this
performance in the 1955-1958 time period. Since the return from this type of
research vehicle diminshes with time, the project will be aimed toward obtaining
a vehicle, not necessarily optimum, which meets the performance requirements
and which will be available for the research program in $3\frac{1}{2}$ years. (Secret)

21.b. Approach

All major airframe contractors have been invited to propose designs in
a competition announced 30 December 1954. The deadline for the submittal
of proposals is 9 May 1955. The proposals will be evaluated and a recommended
technical order of merit will be established. The recommendation, along with
other pertinent information, will be presented to the "Research Airplane
Committee" for the selection of the design which will be developed. The design
approach which has been selected will be presented to the Coordinating Committee
on Piloted Aircraft, Department of Defense, for review and approval. (Conf)

21.c. Subtasks

A task or a project, as required, will be established to develop one of
the four rocket engines being considered to a configuration suitable for this
application. The time available for this task is less than three years and is
considered to be critically short. The rocket engine program will be subjected
to a review immediately upon the selection of the winning airframe design to
determine the availability of the engine in the required configuration. (Conf)

21.d. Other Information

(1) General

The project will be conducted under the guidance of a "Research
Airplane Committee" composed of one representative each from the NACA,
Navy and Air Force. The airplane will be demonstrated by the

DD FORM 613-1
FEB 53
PREVIOUS EDITIONS
OF THIS FORM MAY
BE USED.
SECURITY CLASSIFICATION
SECRET
PAGE 2 OF 4 PAGE
C5-38426

R&D PROJECT CARD
CONTINUATION SHEET

SECRET
SECURITY CLASSIFICATION

SECRET

1. PROJECT TITLE	2. SECURITY OF PROJECT	3. PROJECT NUMBER
(CONFIDENTIAL) X-15 Research Aircraft	SECRET	1226
	4. 1-1226	5. REPORT DATE 7 March 1955

21.d.(1)(contd)

   contractor up to Mach 2.0 at moderate altitude after which the Air Force will conduct a limited Phase II flight test program. After acceptance of the airplane by the Air Force, it will be placed on indefinite loan to the NACA for the flight research program. (Conf)

   (2) Funds

   It is planned to contract for a program which includes mockup, static test and three flight articles. The cost of this program is estimated at a minimum of $25,000M, including engine development costs, over a period of four fiscal years. The Navy is expected to provide one fourth of the total funds required. Funding for any one fiscal year will not exceed $10,000M. A breakdown of R&D funds by fiscal year and an estimate of the man hours required are as follows:

	FY 55	FY 56	FY 57	FY 58	FY 59	
P600	0	$10,000M	$8,000M	$4,000M	$3,000M	
Manhours	10,000	4,000	3,000	3,000	7,000	(Conf)

   (3) Resource Requirements

   (a) A B-36, B-50 or an airplane of comparable size will be required for modification to the carrier configuration.
   (b) All flights of this airplane will be planned for termination at the AFFTC, although on some flights, the airplane may be launched as far away as Salt Lake, Utah.
   (c) Additional instrumentation located remotely from Edwards AFB will be required to monitor and control the flights where remote launching is required. The nature of this instrumentation and its location will be established in the course of the development of the airplane. The requirement for this equipment will probably not occur until 1959, after the initiation of the NACA flight research program. (Conf)

21.e. Background History

   The conception of this airplane seems to have occurred in June 1952 when the Committee on Aerodynamics of the NACA recommended that the NACA increase its research on problems of manned and unmanned flight at altitudes between 12 and 50 miles and at Mach Numbers between 4 and 10. Through a series of studies over

DD FORM 613-1
1 FEB 53
PREVIOUS EDITIONS
OF THIS FORM MAY
BE USED.

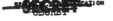

SECURITY CLASSIFICATION
SECRET

PAGE 3 OF 4 PAGES

C5-38426

R&D PROJECT CARD
CONTINUATION SHEET

SECRET
SECURITY CLASSIFICATION

SECRET

1. PROJECT TITLE	2. SECURITY OF PROJECT	3. PROJECT NUMBER
(CONFIDENTIAL) X-15 Research Aircraft	SECRET	1226
	4. 1-1226	5. REPORT DATE 7 March 1955

21.e.(contd)

a two year period conducted independently by NACA's Langley and Ames Laboratories and High-Speed Flight Station, NACA concluded that a new research airplane capable of exploring that flight regime is both necessary and feasible. In May 1954, NACA proposed to the Air Force that a meeting be held to discuss the need for a new research airplane. Concurrent with the NACA consideration of the need for a new research airplane, the Aircraft Panel of the Air Force's Scientific Advisory Board had also been considering the matter and had formally recommended that the Air Force initiate action on such a program. A series of meetings among various elements of NACA, Navy, USAF and Department of Defense resulted in a decision by the Department of Defense that an Air Force managed project under the guidance of a joint NACA, Navy, Air Force Steering Committee would be appropriate. On 6 October 1954, Hq USAF issued Technical Program Requirement No. 1-1 directing the initiation of the project. (Secret)

21.f. Future plans for the development of other Research Airplanes will be contingent upon the results of the X-1A, X-1B, X-1E and X-2 flight programs and the establishment of the need for data in some yet unexplored regime of flight. (Uncl)

21.g. References

    (1) NACA Report, subject, "NACA Views Concerning a New Research Airplane," dated August 1954.

    (2) Hq USAF Technical Program Requirement 1-1 dated 6 October 1954

    (3) Hq ARDC Technical Requirement 54 dated 26 October 1954.   (Uncl)

21.h. Capt. C.E. McCollough, Jr. WCSFF, 39159

UNITED STATES AIR FORCE
HEADQUARTERS AIR RESEARCH & DEVELOPMENT COMMAND
OFFICIAL | AUTHENTICATION

This document is classified SECRET in accordance with par 23C of AFR 205-1.

SECRET

DD FORM 613-1
1 FEB 53

PREVIOUS EDITIONS OF THIS FORM MAY BE USED.

SECURITY CLASSIFICATION
SECRET

PAGE 4 OF 4 PAGES
C5-30426

Appendix 6

X-15 Flight Designation System

The X-15 flight designation system used for the vast majority of the program was formalized in this 24 May 1960 letter from Paul Bikle.

May 24, 1960

From NASA Flight Research Center
To   NASA Headquarters   RSS (Mr. H. Brown)

Subject: X-15 flight designation

    1. At the suggestion of ARDC a system of flight designation for X-15 flight operations has been agreed upon by NASA FRC, AFFTC, and NAA personnel. The system will cover completed flights as well as planned flights; therefore, all personnel concerned should use the flight-designation system as soon as possible.

    2. The flight-designation system consists of a three-column designation. The first column indicates the X-15 airplane by number (1, 2, or 3). The second column indicates the particular free-flight number of a given X-15, or whether the mission was a planned captive flight (C) or an aborted flight (A). The third column indicates the number of airborne X-15/B-52 missions for a given X-15. Designations of flights to date are:

X-15-1	X-15-1 (Cont'd.)	X-15-2	X-15-2 (Cont'd)
1-C-1	1-4-9	2-C-1	2-3-9
1-A-2	1-5-10	2-A-2	2-A-10
1-A-3	1-6-11	2-1-3	2-4-11
1-A-4	1-7-12	2-A-4	2-5-12
1-1-5	1-8-13	2-A-5	2-6-13
1-A-6		2-2-6	2-A-14
1-2-7		2-A-7	2-7-15
1-3-8		2-A-8	2-8-16
			2-A-17

    3. The designation of the next scheduled flights on all X-15 airplanes will be 1-9-14 (X-15-1), 2-9-18 (X-15-2), and 3-1-1 (X-15-3).

Paul F. Bikle
Director, NASA Flight Research Center

TWF:pm

TAT

DEB

Copies to:

  NASA Ames Research Center (2)
  NASA Langley Research Center
    Attention: Mr. H. A. Soule' (3)

Appendix 7

Major Michael J. Adams Joins the Program

DEPARTMENT OF THE AIR FORCE
HEADQUARTERS, AIR FORCE FLIGHT TEST CENTER (AFSC)
EDWARDS AIR FORCE BASE, CALIF. 93523

REPLY TO
ATTN OF: FTTO

SUBJECT: Selection of Crew Member for X-15 Program   1 4 JUL 1966

TO: NASA (Mr. Bickle)

1.   Major Michael J. Adams has been selected from a number of our
experimental test pilots to participate in the X-15 program. His
selection was based on experience and past performance displayed
while assigned to the Air Force Flight Test Center. Major Adams
completed the Experimental Test Pilot Course (Class 62-c) and the
Aerospace Research Pilot Course, graduating number one in his class
from the Experimental Test Pilot Course. While assigned to the
Directorate of Flight Test Operations, he completed a variety of
test projects which included F-5A Category II Stability and Control
Tests and a longitudinal variable stability investigation to determine
optimum fighter aircraft characteristics. We believe Major Adams
has the ability and sound mature judgment required to adapt to the
rigors of a research program such as the X-15.

2.   A brief resume of his military and flight experience follows:

Year	Assignment	Location
1950 - 1951	Enlisted in USAF (Basic Training)	Lackland AFB, Texas
	Link Trainer Instructor	Reese AFB, Texas
1951 - 1952	Aviation Cadet (Pilot Training)	Spencefield, Georgia
		Webb AFB, Texas
1952	Combat Crew Training (F-80/F-86)	Nellis AFB, Nevada
1952 - 1953	Fighter Pilot, 80th FBS (49 Combat Missions)	K-13, Suwon, Korea
1954 - 1956	Fighter Pilot, 613th FBS	England AFB, La.
1956 - 1958	Student (B.S.A.E.)	Univ of Oklahoma
1958 - 1959	Student (Grad Astronautics)	MIT, Cambridge, Mass.
1959 - 1962	Instructor (Maint Offr Course)	Chanute AFB, Ill.
1962 - 1963	Student (Exp Test Pilot Course 62-C)	Edwards AFB, Calif.
1963	Aerospace Research Pilot School Class IV	Edwards AFB, Calif.
1963 - 1965	Experimental Test Pilot (Fighter Branch)	Edwards AFB, Calif.
1965 - 1966	Crew Member (MOL)	SSD, El Segundo, Calif.

Flying Experience:

Total Time	3940:00	
Single Engine Jet	2505:00	(F-80/F-84F/F-86/F-104/F-106/T-33 primarily)
Multi Jet	477:00	(F-5/T-38/F-101 primarily)

3.   As additional information, a photograph and brief biographic sketch are
included.

Clyde B. Cherry

CLYDE B. CHERRY, Colonel, USAF   2 Atch
Deputy for Systems Test   1. Photo
                                        2. Biographical Sketch

                                        Cy to:
                                        ASD (ASZVE)
                                        AFSC (SCSAN/Col Lake)

Major Michael J. Adams was assigned to the X-15 program in the summer of 1966, coming straight from the ill-fated Manned Orbiting Laboratory program.

Adams would make six successful X-15 flights, but was killed during a high altitude flight on 15 November 1967.

Appendix 8

Astronaut Wings

Over the years there has been a great deal of debate regarding if the X-15 pilots were "astronauts." By the definitions in place at the time, the Air Force pilots that flew above 50 statue miles altitude were awarded Astronaut Wings. Under these rules, Adams, Engle, Knight, Rushworth, and White qualified.

The orders that awarded Astronaut Wings to the Air Force pilots were nothing out of the ordinary. A simple sheet of paper—no certificate; not even an embossed seal or a real signature.

Michael Adams was awarded his Astronaut Wings posthumously after he was killed on his only flight above 50 miles. This copy of his orders was largely responsible for getting Adams' name on the Astronaut Memorial at the Kennedy Space Center, Florida.

DEPARTMENT OF THE AIR FORCE
WASHINGTON

AERONAUTICAL ORDER   22 April 1968
40

MAJ WILLIAM J KNIGHT, FR53263, AF Flight Test Center, AFSC, Edwards AFB, Calif 93523, is awarded the aeronautical rating of COMMAND PILOT ASTRONAUT per para 1-22, AFM 35-13. Authority: Para 1-20, AFM 35-13.

BY ORDER OF THE SECRETARY OF THE AIR FORCE

J. P. McCONNELL, General, USAF
Chief of Staff

R. J. PUGH, Colonel, USAF
Director of Administrative Services

DEPARTMENT OF THE AIR FORCE
WASHINGTON

AERONAUTICAL ORDER   15 November 1967
130

MAJ MICHAEL J ADAMS, FR24934, AF Flight Test Center, AFSC, Edwards AFB, Calif 93523, is awarded the aeronautical rating of COMMAND PILOT ASTRONAUT per para 1-22, AFM 35-13. Authority: Para 1-20, AFM 35-13.

BY ORDER OF THE SECRETARY OF THE AIR FORCE

J. P. McCONNELL, General, USAF
Chief of Staff

R. J. PUGH, Colonel, USAF
Director of Administrative Services

Appendix 9

X-15 Program Flight Log

Flight No.	Flight ID	Serial No.	Date	Pilot	Max. Mach	Max. Altitude	Max. Speed
1	1-1-5	56-6670	08 Jun 59	Crossfield	0.79	37,550	522
2	2-1-3	56-6671	17 Sep 59	Crossfield	2.11	52,341	1,393
3	2-2-6	56-6671	17 Oct 59	Crossfield	2.15	61,781	1,419
4	2-3-9	56-6671	05 Nov 59	Crossfield	1.00	45,462	660
5	1-2-7	56-6670	23 Jan 60	Crossfield	2.53	66,844	1,669
6	2-4-11	56-6671	11 Feb 60	Crossfield	2.22	88,116	1,466
7	2-5-12	56-6671	17 Feb 60	Crossfield	1.57	52,640	1,036
8	2-6-13	56-6671	17 Mar 60	Crossfield	2.15	52,640	1,419
9	1-3-8	56-6670	25 Mar 60	Walker	2.00	48,630	1,320
10	2-7-15	56-6671	29 Mar 60	Crossfield	1.96	49,982	1,293
11	2-8-16	56-6671	31 Mar 60	Crossfield	2.03	51,356	1,340
12	1-4-9	56-6670	13 Apr 60	White	1.90	48,000	1,254
13	1-5-10	56-6670	19 Apr 60	Walker	2.56	59,496	1,689
14	1-6-11	56-6670	06 May 60	White	2.20	60,938	1,452
15	1-7-12	56-6670	12 May 60	Walker	3.19	77,882	2,111
16	1-8-13	56-6670	19 May 60	White	2.31	108,997	1,590
17	2-9-18	56-6671	26 May 60	Crossfield	2.20	51,282	1,452
18	1-9-17	56-6670	04 Aug 60	Walker	3.31	78,112	2,195
19	1-10-19	56-6670	12 Aug 60	White	2.52	136,500	1,772
20	1-11-21	56-6670	19 Aug 60	Walker	3.13	75,982	1,986
21	1-12-23	56-6670	10 Sep 60	White	3.23	79,864	2,182
22	1-13-25	56-6670	23 Sep 60	Petersen	1.68	53,043	1,108
23	1-14-27	56-6670	20 Oct 60	Petersen	1.94	53,800	1,280
24	1-15-28	56-6670	28 Oct 60	McKay	2.02	50,700	1,333
25	1-16-29	56-6670	04 Nov 60	Rushworth	1.95	48,900	1,287
26	2-10-21	56-6671	15 Nov 60	Crossfield	2.97	81,200	1,960
27	1-17-30	56-6670	17 Nov 60	Rushworth	1.90	54,750	1,254
28	2-11-22	56-6671	22 Nov 60	Crossfield	2.51	61,900	1,656
29	1-18-31	56-6670	30 Nov 60	Armstrong	1.75	48,840	1,155
30	2-12-23	56-6671	06 Dec 60	Crossfield	2.85	53,374	1,881
31	1-19-32	56-6670	09 Dec 60	Armstrong	1.80	50,095	1,188
32	1-20-35	56-6670	01 Feb 61	McKay	1.88	49,780	1,211
33	1-21-36	56-6670	07 Feb 61	White	3.50	78,150	2,275
34	2-13-26	56-6671	07 Mar 61	White	4.43	77,450	2,905
35	2-14-28	56-6671	30 Mar 61	Walker	3.95	169,600	2,760
36	2-15-29	56-6671	21 Apr 61	White	4.62	105,000	3,074
37	2-16-31	56-6671	25 May 61	Walker	4.95	107,500	3,307
38	2-17-33	56-6671	23 Jun 61	White	5.27	107,700	3,603
39	1-22-37	56-6670	10 Aug 61	Petersen	4.11	78,200	2,735
40	2-18-34	56-6671	12 Sep 61	Walker	5.21	114,300	3,618
41	2-19-35	56-6671	28 Sep 61	Petersen	5.30	101,800	3,600
42	1-23-39	56-6670	04 Oct 61	Rushworth	4.30	78,000	2,830
43	2-20-36	56-6671	11 Oct 61	White	5.21	217,000	3,647

Twelve pilots flew the X-15. Scott Crossfield was first. William Dana was last. Pete Knight went more than 4,500 miles per hour. Joe Walker went more than 67 miles high. Michael Adams died.

The X-15 program is arguably the most successful flight research program ever undertaken by the United States. The 199 flights made by the three research airplanes contributed not only to aeronautical science, but provided many answers the United States needed to get to the Moon during Project Apollo.

Flight number 38 represented the first Mach 5 flight made by any manned aircraft.

Flight number 45 represented the first Mach 6 flight made by any manned aircraft.

Flight number 46 was the first flight for the third X-15.

Flight number 52 set an FAI certified altitude record.

Flight number 53 was the first flight with a dynamic pressure over 2,000 psf.

Flight number 62 set another FAI certified altitude record for class.

Flight number 91 was the highest X-15 flight; 354,200 feet—almost 67 miles high

Flight No.	Flight ID	Serial No.	Date	Pilot	Max. Mach	Max. Altitude	Max. Speed
44	1-24-40	56-6670	17 Oct 61	Walker	5.74	108,600	3,900
45	2-21-37	56-6671	09 Nov 61	White	6.04	101,600	4,093
46	3-1-2	56-6672	20 Dec 61	Armstrong	3.76	81,000	2,502
47	1-25-44	56-6670	10 Jan 62	Petersen	0.97	44,750	645
48	3-2-3	56-6672	17 Jan 62	Armstrong	5.51	133,500	3,765
49	3-3-7	56-6672	05 Apr 62	Armstrong	4.12	180,000	2,850
50	1-26-46	56-6670	19 Apr 62	Walker	5.69	154,000	3,866
51	3-4-8	56-6672	20 Apr 62	Armstrong	5.31	207,500	3,789
52	1-27-48	56-6670	30 Apr 62	Walker	4.94	246,700	3,489
53	2-22-40	56-6671	08 May 62	Rushworth	5.34	70,400	3,524
54	1-28-49	56-6670	22 May 62	Rushworth	5.03	100,400	3,450
55	2-23-43	56-6671	01 Jun 62	White	5.42	132,600	3,675
56	1-29-50	56-6670	07 Jun 62	Walker	5.39	103,600	3,672
57	3-5-9	56-6672	12 Jun 62	White	5.02	184,600	3,517
58	3-6-10	56-6672	21 Jun 62	White	5.08	246,700	3,641
59	1-30-51	56-6670	27 Jun 62	Walker	5.92	123,700	4,104
60	2-24-44	56-6671	29 Jun 62	McKay	4.95	83,200	3,280
61	1-31-52	56-6670	16 Jul 62	Walker	5.37	107,200	3,674
62	3-7-14	56-6672	17 Jul 62	White	5.45	314,750	3,832
63	2-25-45	56-6671	19 Jul 62	McKay	5.18	85,250	3,474
64	1-32-53	56-6670	26 Jul 62	Armstrong	5.74	98,900	3,989
65	3-8-16	56-6672	02 Aug 62	Walker	5.07	144,500	3,438
66	2-26-46	56-6671	08 Aug 62	Rushworth	4.40	90,877	2,943
67	3-9-18	56-6672	14 Aug 62	Walker	5.25	193,600	3,747
68	2-27-47	56-6671	20 Aug 62	Rushworth	5.24	88,900	3,534
69	2-28-48	56-6671	29 Aug 62	Rushworth	5.12	97,200	3,447
70	2-29-50	56-6671	28 Sep 62	McKay	4.22	68,200	2,765
71	3-10-19	56-6672	04 Oct 62	Rushworth	5.17	112,200	3,493
72	2-30-51	56-6671	09 Oct 62	McKay	5.46	130,200	3,716
73	3-11-20	56-6672	23 Oct 62	Rushworth	5.47	134,500	3,716
74	2-31-52	56-6671	09 Nov 62	McKay	1.49	53,950	1,019
75	3-12-22	56-6672	14 Dec 62	White	5.65	141,400	3,742
76	3-13-23	56-6672	20 Dec 62	Walker	5.73	160,400	3,793
77	3-14-24	56-6672	17 Jan 63	Walker	5.47	271,700	3,677
78	1-33-54	56-6670	11 Apr 63	Rushworth	4.25	74,400	2,864
79	3-15-25	56-6672	18 Apr 63	Walker	5.51	92,500	3,770
80	1-34-55	56-6670	25 Apr 63	McKay	5.32	105,500	3,654
81	3-16-26	56-6672	02 May 63	Walker	4.73	209,400	3,488
82	3-17-28	56-6672	14 May 63	Rushworth	5.20	95,600	3,600
83	1-35-56	56-6670	15 May 63	McKay	5.57	124,200	3,856
84	3-18-29	56-6672	29 May 63	Walker	5.52	92,000	3,858
85	3-19-30	56-6672	18 Jun 63	Rushworth	4.97	223,700	3,539
86	1-36-57	56-6670	25 Jun 63	Walker	5.51	111,800	3,911
87	3-20-31	56-6672	27 Jun 63	Rushworth	4.89	285,000	3,425
88	1-37-59	56-6670	09 Jul 63	Walker	5.07	226,400	3,631
89	1-38-61	56-6670	18 Jul 63	Rushworth	5.63	104,800	3,925
90	3-21-32	56-6672	19 Jul 63	Walker	5.50	347,800	3,710
91	3-22-36	56-6672	22 Aug 63	Walker	5.58	354,200	3,794
92	1-39-63	56-6670	07 Oct 63	Engle	4.21	77,800	2,834
93	1-40-64	56-6670	29 Oct 63	Thompson	4.10	74,400	2,712
94	3-23-39	56-6672	07 Nov 63	Rushworth	4.40	82,300	2,925
95	1-41-65	56-6670	14 Nov 63	Engle	4.75	90,800	3,286

Flight No.	Flight ID	Serial No.	Date	Pilot	Max. Mach	Max. Altitude	Max. Speed
96	3-24-41	56-6672	27 Nov 63	Thompson	4.94	89,800	3,310
97	1-42-67	56-6670	05 Dec 63	Rushworth	6.06	101,000	4,018
98	1-43-69	56-6670	08 Jan 64	Engle	5.32	139,900	3,616
99	3-25-42	56-6672	16 Jan 64	Thompson	4.92	71,000	3,242
100	1-44-70	56-6670	28 Jan 64	Rushworth	5.34	107,400	3,618
101	3-26-43	56-6672	19 Feb 64	Thompson	5.29	78,600	3,519
102	3-27-44	56-6672	13 Mar 64	McKay	5.11	76,000	3,392
103	1-45-72	56-6670	27 Mar 64	Rushworth	5.63	101,500	3,827
104	1-46-73	56-6670	08 Apr 64	Engle	5.01	175,000	3,468
105	1-47-74	56-6670	29 Apr 64	Rushworth	5.72	101,600	3,906
106	3-28-47	56-6672	12 May 64	McKay	4.66	72,800	3,084
107	1-48-75	56-6670	19 May 64	Engle	5.02	195,800	3,494
108	3-29-48	56-6672	21 May 64	Thompson	2.90	64,200	1,865
109	2-32-55	56-6671	25 Jun 64	Rushworth	4.59	83,300	3,104
110	1-49-77	56-6670	30 Jun 64	McKay	4.96	99,600	3,334
111	3-30-50	56-6672	08 Jul 64	Engle	5.05	170,400	3,520
112	3-31-52	56-6672	29 Jul 64	Engle	5.38	78,000	3,623
113	3-32-53	56-6672	12 Aug 64	Thompson	5.24	81,200	3,535
114	2-33-56	56-6671	14 Aug 64	Rushworth	5.23	103,300	3,590
115	3-33-54	56-6672	26 Aug 64	McKay	5.65	91,000	3,863
116	3-34-55	56-6672	03 Sep 64	Thompson	5.35	78,600	3,615
117	3-35-57	56-6672	28 Sep 64	Engle	5.59	97,000	3,888
118	2-34-57	56-6671	29 Sep 64	Rushworth	5.20	97,800	3,542
119	1-50-79	56-6670	15 Oct 64	McKay	4.56	84,900	3,048
120	3-36-59	56-6672	30 Oct 64	Thompson	4.66	84,600	3,113
121	2-35-60	56-6671	30 Nov 64	McKay	4.66	87,200	3,089
122	3-37-60	56-6672	09 Dec 64	Thompson	5.42	92,400	3,723
123	1-51-81	56-6670	10 Dec 64	Engle	5.35	113,200	3,675
124	3-38-61	56-6672	22 Dec 64	Rushworth	5.55	81,200	3,593
125	3-39-62	56-6672	13 Jan 65	Thompson	5.48	99,400	3,712
126	3-40-63	56-6672	02 Feb 65	Engle	5.71	98,200	3,885
127	2-36-63	56-6671	17 Feb 65	Rushworth	5.27	95,100	3,539
128	1-52-85	56-6670	26 Feb 65	McKay	5.40	153,600	3,702
129	1-53-86	56-6670	26 Mar 65	Rushworth	5.17	101,900	3,580
130	3-41-64	56-6672	23 Apr 65	Engle	5.48	79,700	3,657
131	2-37-64	56-6671	28 Apr 65	McKay	4.80	92,600	3,260
132	2-38-66	56-6671	18 May 65	McKay	5.17	102,100	3,541
133	1-54-88	56-6670	25 May 65	Thompson	4.87	179,800	3,418
134	3-42-65	56-6672	28 May 65	Engle	5.17	209,600	3,754
135	3-43-66	56-6672	16 Jun 65	Engle	4.69	244,700	3,404
136	1-55-89	56-6670	17 Jun 65	Thompson	5.14	108,500	3,541
137	2-39-70	56-6671	22 Jun 65	McKay	5.64	155,900	3,938
138	3-44-67	56-6672	29 Jun 65	Engle	4.94	280,600	3,432
139	2-40-72	56-6671	08 Jul 65	McKay	5.19	212,600	3,659
140	3-45-69	56-6672	20 Jul 65	Rushworth	5.40	105,400	3,760
141	2-41-73	56-6671	03 Aug 65	Rushworth	5.16	208,700	3,602
142	1-56-93	56-6670	06 Aug 65	Thompson	5.15	103,200	3,534
143	3-46-70	56-6672	10 Aug 65	Engle	5.20	271,000	3,550
144	1-57-96	56-6670	25 Aug 65	Thompson	5.11	214,100	3,604
145	3-47-71	56-6672	26 Aug 65	Rushworth	4.79	239,600	3,372
146	2-42-74	56-6671	02 Sep 65	McKay	5.16	239,800	3,570
147	1-58-97	56-6670	09 Sep 65	Rushworth	5.25	97,200	3,534

Flight number 109 was the first flight of the modified X-15A-2.

Flight number 114 had the nose gear inadvertently extend at Mach 4.2.

Flight number 119 was the first flight with wing-tip pods installed.

Flight number 127 had the right main skid extend inadvertently at Mach 4.3 and 85,000 feet.

Flight number 131 flew with the damper (augmentation) off at a dynamic pressure of 1,500 psf; the highest of the program.

Flight No.	Flight ID	Serial No.	Date	Pilot	Max. Mach	Max. Altitude	Max. Speed
148	3-48-72	56-6672	14 Sep 65	McKay	5.03	239,000	3,519
149	1-59-98	56-6670	22 Sep 65	Rushworth	5.18	100,300	3,550
150	3-49-73	56-6672	28 Sep 65	McKay	5.33	295,600	3,732
151	1-60-99	56-6670	30 Sep 65	Knight	4.06	76,600	2,718
152	3-50-74	56-6672	12 Oct 65	Knight	4.62	94,400	3,108
153	1-61-101	56-6670	14 Oct 65	Engle	5.08	266,500	3,554
154	3-51-75	56-6672	27 Oct 65	McKay	5.06	236,900	3,519
155	2-43-75	56-6671	03 Nov 65	Rushworth	2.31	70,600	1,500
156	1-62-103	56-6670	04 Nov 65	Dana	4.22	80,200	2,765
157	1-63-104	56-6670	06 May 66	McKay	2.21	68,400	1,434
158	2-44-79	56-6671	18 May 66	Rushworth	5.43	99,000	3,689
159	2-45-81	56-6671	01 Jul 66	Rushworth	1.70	44,800	1,061
160	1-64-107	56-6670	12 Jul 66	Knight	5.34	130,000	3,661
161	3-52-78	56-6672	18 Jul 66	Dana	4.71	96,100	3,217
162	2-46-83	56-6671	21 Jul 66	Knight	5.12	192,300	3,568
163	1-65-108	56-6670	28 Jul 66	McKay	5.19	241,800	3,702
164	2-47-84	56-6671	03 Aug 66	Knight	5.03	249,000	3,440
165	3-53-79	56-6672	04 Aug 66	Dana	5.34	132,700	3,693
166	1-66-111	56-6670	11 Aug 66	McKay	5.21	251,000	3,590
167	2-48-85	56-6671	12 Aug 66	Knight	5.02	231,100	3,472
168	3-54-80	56-6672	19 Aug 66	Dana	5.20	178,000	3,607
169	1-67-112	56-6670	25 Aug 66	McKay	5.11	257,500	3,543
170	2-49-86	56-6671	30 Aug 66	Knight	5.21	100,200	3,543
171	1-68-113	56-6670	08 Sep 66	McKay	2.44	73,200	1,602
172	3-55-82	56-6672	14 Sep 66	Dana	5.12	254,200	3,586
173	1-69-116	56-6670	06 Oct 66	Adams	3.00	75,400	1,977
174	3-56-83	56-6672	01 Nov 66	Dana	5.46	306,900	3,750
175	2-50-89	56-6671	18 Nov 66	Knight	6.33	98,900	4,250
176	3-57-86	56-6672	29 Nov 66	Adams	4.65	92,000	3,120
177	1-70-119	56-6670	22 Mar 67	Adams	5.59	133,100	3,822
178	3-58-87	56-6672	26 Apr 67	Dana	1.80	53,400	1,163
179	1-71-121	56-6670	28 Apr 67	Adams	5.44	167,200	3,720
180	2-51-92	56-6671	08 May 67	Knight	4.75	97,600	3,193
181	3-59-89	56-6672	17 May 67	Dana	4.80	71,100	3,177
182	1-72-125	56-6670	15 Jun 67	Adams	5.14	229,300	3,606
183	3-60-90	56-6672	22 Jun 67	Dana	5.34	82,200	3,611
184	1-73-126	56-6670	29 Jun 67	Knight	4.17	173,000	2,870
185	3-61-91	56-6672	20 Jul 67	Dana	5.44	84,300	3,693
186	2-52-96	56-6671	21 Aug 67	Knight	4.94	91,000	3,368
187	3-62-92	56-6672	25 Aug 67	Adams	4.63	84,400	3,115
188	2-53-97	56-6671	03 Oct 67	Knight	6.70	102,100	4,520
189	3-63-94	56-6672	04 Oct 67	Dana	5.53	251,100	3,897
190	3-64-95	56-6672	17 Oct 67	Knight	5.53	280,500	3,869
191	3-65-97	56-6672	15 Nov 67	Adams	5.20	266,000	3,617
192	1-74-130	56-6670	01 Mar 68	Dana	4.36	104,500	2,878
193	1-75-133	56-6670	04 Apr 68	Dana	5.27	187,500	3,610
194	1-76-134	56-6670	26 Apr 68	Knight	5.05	209,600	3,545
195	1-77-136	56-6670	12 Jun 68	Dana	5.15	220,100	3,563
196	1-78-138	56-6670	16 Jul 68	Knight	4.79	221,500	3,382
197	1-79-139	56-6670	21 Aug 68	Dana	5.01	267,500	3,443
198	1-80-140	56-6670	13 Sep 68	Knight	5.37	254,100	3,723
199	1-81-141	56-6670	24 Oct 68	Dana	5.38	255,000	3,716

Flight number 155 was the first flight with (empty) external tanks.

Flight number 166 recorded the highest dynamic pressure of any X-15 flight; 2,202 psf.

Flight number 186 was the first flight with full ablative coating. No tanks.

Flight number 188 was the fastest flight of the X-15 program; 4,520 mph.

Flight number 191 resulted in the death of Major Michael J. Adams; the only fatality during the X-15 program

Index

A

B

F

Faget, Maxim A., 8
Feltz, Charles H., 21
first government X-15 flight, 50
Flight Research Center, 22, 50, 53-54, 62, 72, 79
    See also High-Speed Flight Station
follow-on experiments, 77
    See also Hypersonic Research Engine
Freeman, E. C., Major, 32
fuselage side tunnels, 24, 27

G

Gardner, Trevor, 14
Garrett-AirResearch, 79
General Electric, 13, 38
Gilruth, Robert R., 7
Goldin, Daniel, 81
Greene, Lawrence P., 31
Grumman Aircraft Corporation, 15, 16

H

Haugen, V. R., Colonel, 14
heating projections, 11
Hedgepeth, John, 42
Hermes rocket engine
    See A3 Hermes rocket engine
High Range, 41, 56, 73, 81
high-lift, 10
High-Speed Flight Station, 14, 16, 42, 45, 70
    X-15 proposal evaluation, 17
    See also Flight Research Center
hot-structure, 12, 26-27, 58, 60, 75
Hunley, J. D. "Dill," 67
Hypersonic Research Engine, 57, 62, 79, 80
    dummy ramjet, 58-60

I

Inconel X, 12, 23, 26-27, 29, 31, 54, 58, 75, 79
Intercontinental Ballistic Missile, 8

K

Kelset, Benjamin S., Brigadier General, 14
Kincheloe, Iven C., Captain, 23, 32, 33
Knight, William J. "Pete," Major, 58-59, 61, 63, 67
Kolf, Jack, 60

L

M

N